M·E·A·T

by THE LOBELS

ALPHA BOOKS

NEW YORK, N.Y.

To our father, Morris Lobel,
who founded M. Lobel & Sons,
to our mother, Etta,
and to our brother, Nathan,
who helped us write our book,
we dedicate this book with our love.

The recipes for Beef Stew with Beef (p. 82), Chicken in White Wine (pp. 198–199), Dilled Veal (pp. 111–112), and Oxtail Ragout No. 1 (pp. 86–87) are included in this book by permission of the editors of *Look* magazine. They originally appeared in the April 1, 1969, issue. Copyright © 1969 by Cowles Communications, Inc.

Drawings by Edward Epstein
Typographic design by Judith Lerner

ISBN: 0-944392-05-9

PRODUCED BY MADISON PUBLISHING ASSOCIATES

Manufactured in the United States of America

Contents

Introduction by Marylin Bender   iv

1 ╱ Some Facts about the Meat You Buy   1
2 ╱ How to Care for Meat and Poultry after
      You Buy It   6
3 ╱ How to Carve Meat and Poultry   12
4 ╱ Beef   29
5 ╱ Veal   90
6 ╱ Lamb   115
7 ╱ All about Pork—Yesterday and Today   146
8 ╱ Poultry   177
9 ╱ Variety Meats   231
10 ╱ Game—Wild Birds and Animals   252
11 ╱ Poultry and Meat Stuffings   271
12 ╱ Sauces   282
      Index   297

Introduction

THE OBVIOUS inclination is to call M. Lobel & Sons the Tiffany of Madison Avenue butchers. Or is it the reverse? Perhaps Tiffany is the Lobel of Fifth Avenue jewelers. On further reflection, a more accurate comparison would be between these perfectionist meat men and Fulcro di Verdura, the cognoscenti's *bijoutier,* who, incidentally, is a Lobel customer and who has contributed a recipe to this cookbook.

For although the Lobels ply their trade in a showcase shop on Manhattan's upper East Side in full sight of the strolling and bus-riding throng, and friendly and unsnobbish though their manner be, they can by no stretch of the term be called everyone's butcher. Neither their prices nor the quality of their meats are scaled to mass acceptance. There are never any bargains or discounts or seasonal specials at Lobel. Only connoisseurs can afford Lobel prices, which is to say only those who believe that the finest ingredients are the deep secret of haute cuisine.

Such connoisseurs include the rich and the petty bourgeois, the show-offs and the shy, those once dubbed The Beautiful People as well as a host of anonymous gourmets and gluttons. The Lobels supply the venison for a multimillionaire's dinner party and the pot roast that his cook serves to her friends on her night off. Then there are the rest in between, who— once introduced to cooking with prime meat, like Eve after her bite of Eden Delicious—could simply never return to the innocence of simple economy.

It's tempting to talk about the Lobels in terms of their customers, as if they had an effect on the shop or its products, which they do not. On the other hand, it must be said that the Lobel fame ripped beyond the inner circle after Anne-Marie Huste was fired by Jacqueline Kennedy. Miss Huste was the pretty, young German cook whose weight-watching recipes appeared in a magazine with the implication that they were responsible for the former First Lady's razor-blade chic. The indiscretion cost the cook her job but won her secondary-celebrity status and a chance to divulge other secrets, such as where she shopped for Mrs. Kennedy (and for herself). Actually, Mrs. Kennedy had discovered the Lobel shop, which is at Madison Avenue and 82nd Street, three blocks from her Fifth Avenue apartment, and had recommended it to her cook. After the contretemps the Lobels continued to supply both the President's widow and her former employee, who became a radio and television personality and a cookbook writer.

When Mrs. Kennedy married Aristotle Onassis, Suzy Knickerbocker, the society gossip columnist, reported that Stanley Lobel had flown to Athens to supervise the cutting of the meat for the nuptial banquet. Stanley denied this bit of patent nonsense. Or is it? Although Stanley did not, in fact, go to Greece for Onassis, his brother, Leon, has made one-day round-trip journeys to Fort Lauderdale, Florida, with a suitcase packed with steaks for another tycoon's parties.

The Lobels perform other services for customers, which inevitably led to the suggestion that this book be written. There are the evening seminars on meat anatomy and preparation for groups of housewives and the spontaneous cooking tips and recipes delivered over the counter or by telephone. Many New Yorkers, particularly of the rich and famous genre, do all of their shopping by mail and phone, and it is not unusual for the Lobels to receive an emergency call from a celebrated personage on cook's day off (or on the day cook quit), plaintively reporting that the lamb chops don't seem to be getting done as they should. The Lobels' first diagnostic query: "Are you sure you turned the broiler on?"

One world-renowned beauty whose most pedestrian activities rate fan-magazine headlines had the Lobels acting as her secret chefs during one tense week when she was between cooks and in the market for her next husband. They prepared the soup and meat courses. Dessert came from the neighborhood fruit stand and French bakery. For others with slightly greater culinary competence, the crises have to do with a discrepancy between cookbooks or within a single cookbook recipe.

For still others, the Lobels have been a more comforting source of kitchen counsel than some more-renowned authorities who do not acknowledge the limits of time that a hard-pressed working girl, the mother of several, or a hungry bachelor can spend in food preparation. The Lobels are haughty and unyielding about only one ingredient in a recipe —the meat—which they insist must be of top quality, and if not, *caveat emptor*. And *caveat* cook.

Yet the Lobels are so tolerant of personal idiosyncrasies about food that they have never balked at one customer's order for fresh wild pheasant, boned and chopped for pheasant-burgers. Or at another's eighth of a pound of ground round for her son's turtles. Nor would they ever blink an eyelash at the frequent order (in Manhattan overindulgent dog-owning circles) for porterhouse ground for the canine of the house.

But enough said about their customers. Who are the Lobels?

They are Leon and Stanley, a youngish, wholesomely nice-looking and extraordinarily genial pair of brothers who are doing what their ancestors have done for four generations here and in Europe. In the 1840's an Austrian farmer named Lobel raised cattle and sold them after slaughter to buyers from far and near. His son became a purveyor of meat in another district. His son, Morris Lobel, started out on his own in the early 1900's, buying cattle and slaughtering them for sale to discriminating housewives. In 1911 he pulled up stakes for the United States, where he settled in Boston. There he prospered as a wholesaler and retailer of prime-quality meat for restaurants, hotels, and households.

During the influenza epidemic of 1917 he fell ill and was forced to close his business. When he recovered, he uprooted himself once more, this time for New York, where he established himself first as a wholesaler, then as a retailer, in various sections of the city. In the 1950's he followed his clientele to the fashionable upper East Side of Manhattan, where he found the present store at 1096 Madison Avenue, near 82nd Street. By then he was accompanied by his three sons—Leon, Nathan, and Stanley—who had discarded earlier interests in dentistry and electronics for the family trade. Morris died in 1967; Nathan three years later.

Leon and Stanley carry on the family tradition in what is surely one of the most striking butcher shops of the city. It is a compact shop, gleaming in white enamel, sparkling glass, and stainless steel. The two brothers work from 8 A.M. to 6 P.M., Monday through Saturday, cheerful and tireless against a background of FM radio music and the incessant ring of three telephones and under the gaze of father Morris, enshrined in a miniature painting above their heads and flanked by the horns of the deer and elk he shot in his huntsman's days.

What has made the Lobel shop a Madison Avenue landmark is the windowed cooling room that fronts on the street. A freezer and another cooling room for moist aging are behind the scenes, but the visible refrigerator, set at 33 to 36 degrees, is for dry aging, which seals in the juices of prime steaks and beef roasts. In public view at all times are a trio of steers, six dozen or more shell strips of beef, which are aged for six weeks, and perhaps a lamb that will stay for a week.

In the shop an assistant stands at a block, cleaning the chickens—exquisite birds with skin as white and gold as sun-kissed babies. Lobel chickens are raised by a New Jersey farmer as though they were beauty-contest winners. They are fed on corn and sheltered in raised cages so that barnyard scraps shall never cross their beaks. There are Long Island ducklings and, at Christmas, geese that are nurtured in a freshwater pond on grain and corn; quail and pheasant from private preserves; baby veal of a milk-white complexion seldom seen on this side of the Atlantic; venison in fall and lamb for Easter; calves' liver with a lucid skin that slices like butter after a few minutes in a sauté pan; fresh hams that have traveled to Boars Head for smoking and returned ("Everything we sell goes through our hands fresh," says Leon Lobel).

Displayed like kings' ransoms, they leave the shop in the style befitting their caste: white-paper leggings on the crown roasts, roasts with their rind frilled, then wrapped in white, the steaks with their freezer paper lacking only a gold ribbon to go beneath the Christmas tree. And how many pounds of Lobel meats do go out as presents for those who have everything else? The shell steaks cut butterfly-style and heart-shaped for Valentine's Day, the boneless beef roasts and smoked geese for winter bouquets, the elegant sirloins in the Steak of the Month Club, which brings Christmas twelve times a year.

To assemble these delicacies the Lobel brothers lead a stringent life, early to bed and early to rise. They are up at 4:30 A.M. to be at the wholesale meat market in the Bronx at 5 A.M. to roam through the lanes and alleys of stalls for three hours, scrutinizing the herds of beef and flocks of lambs, judging color and configuration, tagging their choices in yellow and red, and finally settling their accounts at 7 A.M. in time to drive south into Manhattan and open shop at 8.

Glorious ingredients are nine-tenths of a feast, but some knowledge of basic food preparation goes a long way for the other tenth. In this book, which tells all you should know about meat from the moment you see it in the butcher's display case until you serve it for dinner, the Lobels share their lore as well as some recipes collected from the Lobel recipe library and celebrity customers. *Bon appétit!*

MARYLIN BENDER

Some Facts about the Meat You Buy

DO YOU EVER think about shopping for a nice juicy steak at five o'clock in the morning? Well, your butcher does. He's thinking about your daily meat requirements while you are happily snoozing.

In the early hours of the morning, the wholesale meat market is buzzing. All during the night, carloads of meat arrive from livestock farms, from slaughterhouses. Carcasses of beef, veal, lamb, pork—from Kansas . . . from Illinois . . . from Iowa . . . from Missouri . . . from all over the country— come rolling in, packed in refrigerated trains and trailers. Then they are hung on display in the packing center for butchers to examine and pick and choose.

The wholesale market in New York City occupies many city blocks. To keep the meat fresh, the building interiors are kept at a year-round temperature of from 32° to 38° F. It takes a conscientious butcher a good three hours to cover the scene thoroughly, and you can be sure that he is well equipped with insulated or warm clothing.

The rows of carcasses are hung so close together that there is just enough room for the butchers to walk by. We look until we find a steer that is perfect for our Prime market. The animal's conformation will reveal its age, which should be from ten to fourteen months. The flank must feel thick, the bones be blood red, and there must be an outside layer of firm, creamy fat. The final test is a good look at the club "eye" or "face." This is in the center of the animal (where club steaks come from). Then we continue walking and watching to equal our find in baby veal and young lamb and pork.

Through these wholesale meat markets, 35 billion pounds of beef, veal, pork, and lamb are processed each year, the number of pounds that Americans consume. And it amounts to more tonnage than that of all the American automobiles produced annually.

The popularity of meat seems to be steadily increasing. In the past thirty-five years, the per capita intake has increased more than one pound each year in the United States and now Americans eat more than 175 pounds of meat per person per year.

In past years, not only has the quantity of meat increased but the quality has improved. Breeding has become more scientific. Methods of animal care, feeding, and slaughtering have been improved. There have been innovations in processing and refrigeration techniques, and improved transportation from farms all over the country to packing centers or wholesale meat markets.

The U.S. Government has become more stringent. Inspection of the condition of live animals and carcasses has been stepped up. Each wholesale cut of an inspected and approved carcass must have federal stamps like these:

These purple stamps are a guarantee that the meat came from healthy animals and that such meat was processed under sanitary conditions. All fresh and processed meat products that are shipped from one state to another must have these federal stamps. Every butcher automatically looks for them because they tell him that the meat is wholesome, although they tell him nothing about quality or grades of meat. The U.S. Department of Agriculture is willing to provide grade stamps to any meatpacker who wishes to pay a fee. The wholesomeness stamps are mandatory and are free. The grade stamps are optional, but many packers find that they are helpful in their dealings with less knowledgeable butchers. These stamps also give retail customers a guide to meat quality. If you are dealing with a new butcher, you can ask him what grade of meat you are getting, and more often than not (if he wants your future patronage) he will show you the grade stamp.

There are three grades of meat you will encounter in your shopping. Prime is the highest grade of beef, veal, and lamb and is marked with a purple shield-type roller stamp like this:

This grade stamp is rolled on in a ribbonlike print along the length of the carcass and across the shoulders. Prime meat is usually found only in smaller meat shops. Much of the wholesale prime meat is bought by top restaurants and hotels. But if you are looking for a bargain in less expensive cuts, don't feel that you might just as well buy them from lesser-grade animals. The quality of prime meat shows up in any section of the animal's carcass, so if you want to experiment with less expensive cuts it is still wiser to buy prime quality—you will realize the difference in flavor and texture, whether it is in a pot roast, a stew, or a casserole. Even in such variety cuts as oxtails, sweetbreads, or calves' brains, the quality of the meat will affect the taste of the final dish.

The next grade after prime is choice. Then comes the

grade that is called good. These are marked with the same type of USDA purple shield and are the grades that usually appear in large supermarkets. All of these grades are wholesome; the variation is in their tenderness, juiciness, and taste. The lower grades are: standard, commercial, utility, cutter, canner (in beef); standard, utility, cull (in veal); utility, cull (in lamb). These are sold to canners and go into such processed meat items as hot dogs, salami, bologna, and so on.

Instead of using the top three federal grade stamps, many packers use brand names to indicate the quality levels of their products. Your butcher will be aware of what each name means, although you may not. It is important therefore to deal with a butcher you can trust. With a good customer-butcher relationship, you will get honest service.

"Well aged" is a term that is something of a mystery to most consumers. Many people know about the length of time some sportsmen like to hang the game they have shot (until it falls off the limb of the tree where they have strung it). But domestic meat is quite different. It needs only a limited amount of aging. For example, we feel that beef should hang in our coolers for three to four weeks (lamb and veal about one week). This is what is called dry aging—until a thin coating covers the carcass, which will then retain its flavors and valuable juices.

The tremendous appetites of Americans for meat is no secret—nor that they are often disappointed by the meat available in foreign countries. American meat is the very best that can be had. In fact, the S.S. *France,* which is noted for its excellent cuisine, buys all of its meat while docked in New York.

As with meat, there are federal grading stamps that are available to poultry farmers who wish to pay the fee. These are also shield-type stamps and go from A to B to C (see Chapter 5). The A grade means that the bird is young, tender, well fleshed with a good cover of fat, and has a perfect conformation. Grades B and C are imperfect. For example, in a turkey the breastbone may be dented, curved, or have a crooked conformation; or it may be poorly fleshed (scrawny) and lacking in fat covering.

Poultry is raised in practically every country of the world, and when Americans go abroad they frequently select a poultry dish to sample unusual ways of preparing this food.

Even though the United States has the best-quality meat in the world, some other countries are noted for their special methods of achieving meat delicacies. For example, American farmers usually allow their animals to mature with scientific care and feeding. But in France a calf is frequently taken away from the mother at an early age. The calf has fed only on mother's milk and becomes baby veal, with its soft bones and tender white meat. The Italians also slaughter animals that have fed only on mother's milk—in this case the hothouse lambs and suckling pigs that may be available on special order from certain markets in the United States. But they are, of course, quite expensive.

In the following chapters you will find out more about what to look for in meat and poultry, about the proper coloration of meat and fat, about marbling and texture. And you will understand how important such things are in buying meat for your table.

How to Care for Meat and Poultry after You Buy It

THERE ARE definite ways to store fresh meat or fowl in your refrigerator if it is to be cooked in the near future. It is also possible to preserve meat in your freezer. The way you coddle the meat you take home is as important as the way we take care of it before you buy it.

↑ STORING MEAT AND POULTRY

The most important element is to allow as little time as possible to elapse between the butcher shop and your refrigerator or freezer.

FRESH MEAT

This should be stored in your refrigerator in a loose wrapping. The original paper it comes in is best. Just loosen up the ends so that the meat or poultry can "breathe." Don't re-package it in a plastic-type wrapper, as this will decrease the partial drying on the surface that increases the meat's keeping quality. Roasts, steaks, and chops can be kept this way for two to four days in a refrigerator that has a temperature of 38° to

40° F. However, chopped meats and variety meats are much more perishable and should be cooked in one to two days.

FROZEN MEAT AND FOWL

These should go right into the freezer the minute you get them home. But should you buy some frozen foods to take out to the country for a weekend, wrap them in layers of newspaper. This is a great insulator, and the meat should remain stiffly frozen for four to six hours. But back to your home refrigerator. The National Live Stock people have given us the following information: The ice-cube compartment of a refrigerator is not intended as a substitute for a regular freezer. Therefore meat should not be kept there for over a week. A freezer should not go above the temperature of 0° F. This is much lower than the temperature required to make ice cubes.

CURED AND SMOKED MEATS

These, as well as sausages and all ready-to-serve meats, should be refrigerated in their original wrappings. But do not be too optimistic about their keeping powers. Try to use them within a week. Bacon that has been opened should be used up in about two to three days.

CANNED MEATS

These can be kept in their airtight containers for many months—just as they are kept on a grocer's shelf. But once cans are opened they should be refrigerated and used within two days. The only exception is canned ham, in which preservatives are used. But the best canned hams—the kind we sell—are packed with lesser amounts of preservatives and therefore must be stored in the butcher's refrigerator. You must also refrigerate them when you get them home. Be sure to follow the instructions on the outside of the can.

LEFTOVER MEATS

Cooked-meat leftovers should be cooled quickly, then covered or wrapped promptly and placed in the refrigerator. Cooked meat keeps its quality best if it is left in large pieces, even if the bones are removed. Small sections dry up faster. But leftover meat (or gravy or broth) should not be refrigerated for more than one or two days.

⸱ HOME FREEZING

Insist upon the freshest possible meat for home freezing. Then ask your butcher to cut it into the portions you will use. When you get the meat home, wrap it in moisture-vaporproof plastic. Be sure that each piece is wrapped tightly so that the air is sealed out and the moisture is locked in. If air penetrates the package, moisture will be drawn from the surface of the meat and a condition develops that is known as "freezer burn."

Before wrapping any poultry, wash it well and then dry it. Here is a tip for either whole birds or pieces. Season them before you wrap them. Use the type of seasoning you would ordinarily use at cooking time—salt, pepper, even a bit of garlic rubbed over the skin and meat. And if you like the idea of "scrubbing" a bird (especially duck) with wedges of lemon, do that before freezing, inside and out. You will find that such pre-freezing seasonings will not only flavor the skin and cavity but will seep into the meat.

Individual pieces of fowl are wrapped just as meat is—in moisture-vaporproof plastic, but whole birds need slightly different attention. Giblets should be removed from the cavity and frozen separately. They may later be used for stock, gravy, or dressing. After seasoning the fowl inside and out, bend the wings by twisting the wing tips over the forewings. Next, push the legs straight back so that they do not take up extra freezer space. The result is a bird that has a compact, ball-like shape.

We think the best covering for a large bird is a plastic

bag. And here is a trick about extracting the air after you have put the bird inside. Make a small circle with your thumb and index finger, then pull the open end of the bag through this circle. Now all you have to do is place your mouth over the small opening and suck out the air (just the opposite of blowing up a paper bag). You will know when all the air is extracted, as the plastic will cling to the fowl. Then quickly tie the opening with string before any air can reenter.

STORAGE TIME FOR FREEZING

For any home freezing, the temperature should be 0° F. or below. But even under these conditions there are variations in the "keeping" power. For example, beef can stand the waiting test more than most meats. It is quite safe to "hold" beef in the freezer for six months. Lamb and veal are a little more anxious to get out of that freezer compartment—three to six months. But pork resists its stay in the freezer even more—two to three months.

Any type of poultry can be kept frozen for five to six months, but no longer. After this period the quality deteriorates and the skin and meat begin to dry out.

In the case of variety meats, sweetbreads, brains, and liver are so delicate that they have a tendency to dry out if kept frozen for more than one month. Veal and lamb kidneys can stay for two months, and giblets can hold out for about three months.

Leftover meat may be frozen and stored for about one month. After the meal is finished and the meat has cooled, it should be wrapped in the usual freezing method and popped into the freezer the same day.

Every package of meat or poultry to be frozen should be labeled, with the date of storage noted. It is also a good idea to mark down the type of cut and the poundage. In this way you will know immediately when you dig into your freezer "store" what will fit into your menu and how many people a given piece will serve.

An important word of caution: Never, never refreeze meat that has been even slightly thawed.

⁄ HOW TO DEFROST MEAT AND POULTRY

A freezer full of meat and foul can make you feel secure about planning a dinner. But you must "order" from your freezer as if buying fresh meat from a butcher. All meat should be at room temperature before cooking. Therefore you will have to thaw the meat well. The length of defrosting time depends upon the size of the meat.

SMALL CHOPS OR CHICKEN SECTIONS

These can be placed on the lower rack of the refrigerator on the morning of the dinner. About four hours before cooking time, they should be put on the kitchen counter.

LARGE STEAKS

A 2-inch porterhouse should be put on the lowest rack of your refrigerator at dinnertime the night before. If it refuses to defreeze when you poke it in the morning, put it on your kitchen counter at about noon.

ROASTS

Allow a large roast to "rest" in your refrigerator for a couple of days. Then bring it to room temperature early on the morning before roasting.

LARGE TURKEYS

These frozen birds take considerable time to thaw out. It is best to place them in the sink, complete with freezer wrappings, the night before. Immerse birds in tap water and put a towel on top. This cannot be done with a roast, since it does not have the skin covering of the turkey.

SMALL BIRDS

Depending on the size, these should have the same defrosting treatment as chicken, chops, or steak.

⸜ TENDERIZING

There are many methods used by butchers to make lower quality meats seem more tender. Pounding gives them the illusion of tenderness, but sometimes they use tenderizing products that break down the enzymes in the meat. If these products work so efficiently on meat, how do they work on our own insides?

How to Carve Meat and Poultry

CARVING IS TRULY AN ART. But carving the way our grandfathers and great-grandfathers did it has almost become a lost art. In the old days a Thanksgiving- or Christmas-dinner table was not only laden with scrumptious dishes, it was surrounded with eager faces, all turned toward the man who performed wizardry with a knife on a turkey.

First he honed his large carving knife against steel. Then he poked a streamlined fork into the bird to secure it and made incisions to separate the joints of the thighs and the legs. Then, to tantalize the group, he would hone again his already sharp knife, after which serious carving started—thin and even slices, deftly cut. First the breast meat—a little outer crisp skin saved for special people—then the dark meat from the thigh. The legs were up for bids among the youngsters. And so was the tip end of the bird (sometimes called "the-last-thing-to-go-over-the-fence"). The carver usually saved two succulent bits for his wife: the oysters—so called because of their shape—that nestle in the spoon-shaped hollows of the backbone above the hip joints.

It is a ritual that should not be replaced by dinner served at separate tables, or a serve-yourself supper buffet, both of which usually involve kitchen carving—sometimes done by a caterer. Even if you can afford to hire a caterer for large parties, it is well to know just how to carve any type of meat or fowl—whether you do it in the privacy of your kitchen or for "show" at your own dining table.

⸕ CARVING TIPS

There are certain fundamentals that should be kept in mind when you are carving:

- All knives should be made of high-quality steel and be extremely sharp. (A dull knife will not give thin, even slices and will shred the meat.)
- Use knives of the correct size for every carving project.
- After the first incision has been made, the angle of the knife should never be changed.
- For smooth, even slices, each cut should be direct and sharp, with long, sweeping strokes. (A sawlike motion will only result in jagged, uneven cutting.)
- In carving, neatness does not require strength, only sharp tools.
- Cutting against the grain is the usual rule in carving all meat and poultry.

⸕ CARVING IMPLEMENTS

Knives, of course, are a carver's best friends, but forks are important too. Carving sets are favorite wedding presents, but if you do not have such in your trousseau wardrobe, you can buy the needed pieces separately. Sets are fun to own, but you really don't need a special fork for each cutting job.

BONING KNIFE

The sharp blade of this handy little knife is just about 5 inches long. It is slim and tapered and has a sturdy handle.

POULTRY SHEARS

If you are particularly fond of fowl, you will want to own a pair of these. They look a bit like garden shears but are chrome-plated and have sharp serrated edges. They make the carving of small birds simple because they can snip skin, meat, and even bone with ease.

STANDARD CARVING SET

The knife in this three-piece set has a curved blade that is about 9 inches long. The fork has a guard that keeps it from entering the meat too deeply. But the most important part of this trio is the steel, which will keep your knives sharp. This set is basic and can be used to carve almost anything.

STEAK SET

This is a junior edition of the standard and is great for

young-marrieds. The blade is only about 7 inches long, but it has a most attractive look when you are slicing steak at your own table. This set consists of just the sharp little knife and a handy fork.

ROAST CARVER AND CARVER'S HELPER

Although these are not usually sold as a set, they work as a good team. The slicer has a long, flexible blade that may measure from 11 to 16 inches. This pliable knife is marvelous for standing rib roasts and large hams. As you can see, the carver's helper has widely spread tines, designed to hold a large hunk of meat in a very steady position.

✓ SERVING DISHES

When you are carving a large bird or roast, it is important that you have a large carving platter, with room enough on the side to stack the slices you have cut. The platter should not be cluttered with garnishes or accompaniments, as these will only distract the serious business of carving. The carving platter should be quite hot before it receives the bird or roast. And remember that the dinner plates should be as hot when accepting a slice of meat.

✓ SHARPENING KNIVES

Wagons used to drop by country farms with full equipment for knife-sharpening. If memory serves, it was done with moving wheels of sandstone with a bit of water added. Now there are all types of citified sharpeners. There are the roll-it varieties that fit onto your kitchen wall, and there are electric ones that give a sharp edge in a second or two. But knife-sharpeners who are really diehards insist that the best way to hone a knife is with steel. Some prefer to hold the steel end-down on the table and slither the knife around so the blade gets action from every angle. Others prefer to use the basic method, in which the steel is held in one hand while you "slice" the knife back and forth until all of the blade has been sharpened.

CARE OF CUTLERY

Once you have bought the best quality of knives and have honed them to perfection it's up to you to keep them in prime condition. They should not be kept in a drawer with other cutlery where their edges can be snagged. They should be treated gently—as a surgeon treats his most important tools. They may be hung on a metallic board, but the best way is to hide them away in a felt-lined drawer.

⸍ CARVING ROASTS AND HAMS

It takes a real expert to carve these efficiently and well, and efficiency in carving adds up to real economy. For example, a large roast that is ample for two meals can be hacked so that the second meal of cold cuts disappears, or at most becomes a few meager luncheon sandwiches. Once you have mastered the trick of carving roasts and hams, it will seem as easy as slicing an apple pie.

STANDING RIB ROAST OF BEEF

Although it is cooked standing up, this roast is carved face-down. Remove the wedge-shaped slice from large end so that it will lie flat on the platter.

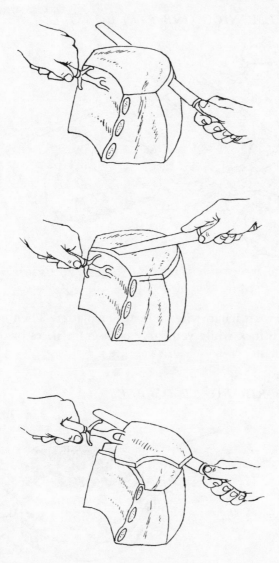

1. Insert fork below top rib bone. Then carve across the face from fat edge to bone. Cut even slices that are ¼ inch or more.
2. Cut along rib bone with tip of knife to release slice.
3. Slide knife back under slice and, steadying it with fork, lift slice to side of platter or hot side plate.

BEEF DELMONICO (RIB EYE) ROAST

Place roast lengthwise on board or platter. Then hold roast firmly with fork while you carve cross-grain slices of the thickness you wish.

ROLLED RIB ROAST OF BEEF

Place roast on platter with the cut surface down. Insert fork, guard up, an inch or two from the top. Slice cross-grain. Remove each cord only as you approach it. Cut cord with tip of knife, loosen it with fork, and lay to one side.

OTHER ROLLED ROASTS

These are placed lengthwise on a platter and are carved like beef Delmonico.

BLADE BONE POT ROAST OF BEEF

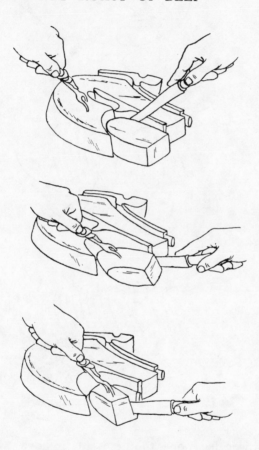

1. Use small carving knife and cut between muscles and around bones to remove one solid section of meat.
2. Turn the section you have removed so that meat fibers are parallel to platter.
3. Hold meat with fork and carve ¼-inch slices cross-grain.

ROAST LEG OF LAMB

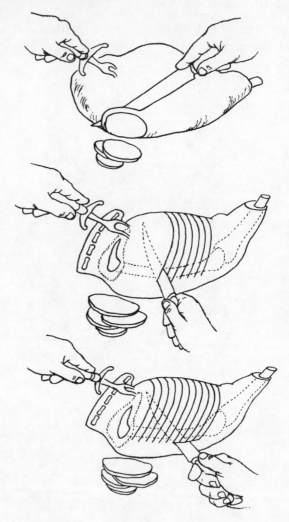

1. With lower leg bone to the right, remove two or three lengthwise slices from thin side of leg, where the kneecap is.
2. Turn roast so that flattened side hugs platter. Starting where shank joins the leg, make slices perpendicular to leg bone.
3. Then loosen slices closely along top of leg bone so that they can be served easily.

LOIN ROAST OF PORK

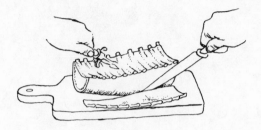

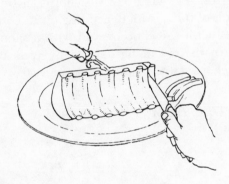

1. The backbone must be removed before the roast is brought to the table. This should be done so that no meat is lost with the bone.

2. Then the roast is placed on the platter with the rib side facing the carver so that he can see the angle and slice accurately. The fork is placed on top of the roast to keep it steady. Slices are now made by cutting closely along each side of the rib bone. One slice will contain the rib, the next will be boneless.

CROWN ROASTS

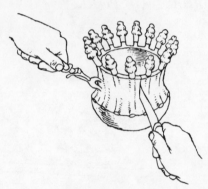

Even though these may look complicated to carve, they are really quite simple. Your butcher has made things easy for you by removing the backbone. You merely slice down between the ribs, removing one rib chop at a time. The stuffing in the center of the crown, depending on its consistency, may be either sliced with the chop, served with a spoon, or sliced like a pie. Crown roasts can be devised by your butcher from a row of rib chops, either lamb, pork, or veal.

WHOLE HAM

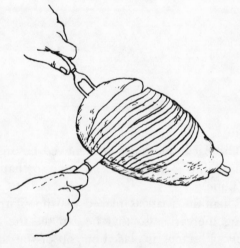

This is a sketch of a perfectly sliced whole ham. The carving procedure is almost identical to that of roast leg of lamb, so follow those instructions.

SHANK HALF OF HAM

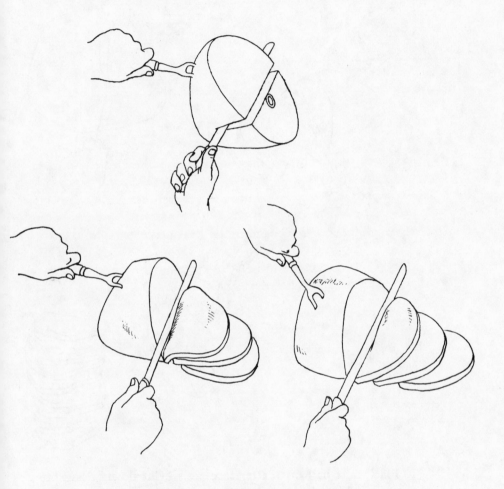

1. With shank at carver's left, turn ham so thick "cushion" side is up. Then cut along top of leg and shank bones.
2. Place cushion meat on carving board with fat side up, then make perpendicular slices.
3. With a sharp boning knife, cut around leg bone and remove all meat. Then turn meat so that thickest side is down and slice.

BUTT HALF OF HAM

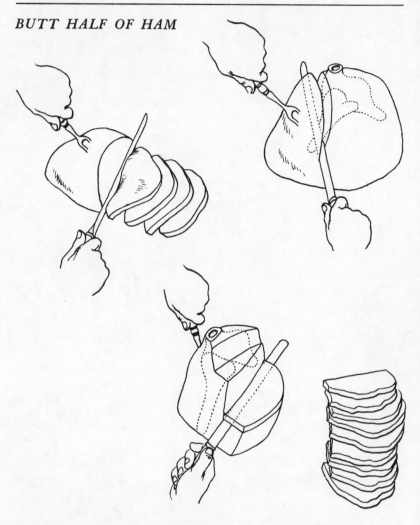

1. Place butt half on platter face-down. Cut down along bone to remove large boneless section.
2. Boneless section is placed on fresh platter and carved in cross-grain slices.
3. Remaining section of butt is now secured by fork on bone side, with meaty side at right of carver. Now slices are cut across meat until knife strikes bone. Each slice should be released from bone with tip of knife, then lifted to side of platter.

⸙ CARVING STEAKS

As with most meat, steaks should be cut against the grain. And if a bone is involved, it should be removed expertly before the cutting process begins. But there are exceptions. If a member of the family is particularly fond of cutting around his steak bone, let him do his own carving at the table. If you wish to be an expert steak-carver, here is the classic method.

T-BONE OR PORTERHOUSE STEAKS

1. Use a small carving knife and fork, or even a boning knife. Place the steak on a platter or board with tail of steak at carver's left. The knife you use must be extremely sharp because you must cut around the T-bone so exactly that it is meatless when you discard it.
2. Now you have solid meat, with no bone to interfere with your carving. You then cut right across the steak, making uniform, wedge-shaped portions. Then the tail is cut, as indicated, into diagonal slices. All of this makes serving extremely easy—each person can have a piece of tenderloin, a piece of the larger section, and a bit of the flank (or tail). In this way, no guest is favored with a particular juicy bit.

/ CARVING POULTRY

Very large birds are the only problems that a host-carver has to face. Very small varieties are up to the individual guests —you only have to supply sharp knives. But even guests should have some knowledge of the anatomy of fowl—where the bone sockets are located and so on—so that they can do a neat at-table job on their personal plates. Because the anatomy of birds— small or large—is so similar, it might be well for a beginner at fowl-carving to start with a medium-sized bird before he goes on to the large birds.

A simple way to study this kind of carving is to buy a chicken and tell your butcher that you wish to cook it in separate parts or that you want it for a fricassee. Then tell him that you would like to see how he dissects it. It's best to go to the shop at some time other than the rush hour, but almost any butcher will be happy to give you a private show. Your main interest should be just how to sever the wings, legs, and thighs —all of the joints. You will find that when the fowl has been well roasted, the separation will be simpler than when your butcher worked on a uncooked bird. Sometimes the joints will come apart with just a twist or two.

Since all poultry is so similar in structure, we will give you the basics for carving a turkey, and you can use these instructions on practically any large bird. Smaller varieties and ducklings yield well to poultry shears.

TURKEY AND LARGE FOWL

There are two classic ways to carve turkey—the standard "upright" style and the "side" style. The latter is used when the bird is so large and the company so small that a second meal is indicated. The carving methods are identical, except that in the side style only half of the turkey is used. Here is the upright style:

1. Remove both legs, including drumstick and thigh. The simple way to do this is to hold the drumstick firmly with your

fingers and pull away from the body, then twist a bit so that the joint loosens. Now cut through the skin between leg and turkey. It will pull freely away from the body. The joint connecting the leg to the backbone will usually snap free. If not, it can easily be severed with a knife point.

2. Now that the complete leg has been severed, carve the dark meat. Use a separate platter for this job and look at the sketch to see how to carve a drumstick.

3. The thigh, of course, has more dark meat than the leg, and it is tender and delicious. Hold this section firmly on a platter and then cut even slices parallel to the bone.

4. Now we come to the white meat. The important thing is to make smooth, even slices. First, you place a knife parallel and as close to the wing as possible. Then you make a deep cut into the breast, right to the bone. This is your "base" cut. All breast slices will stop at this vertical cut.

5. After that good base cut, you can begin slicing with the greatest of ease. You should start carving halfway up the side of the breast, slicing down and ending at the base cut. Of course, each new slice should start slightly higher on the breast.

And remember: Don't carve more than your guests can absorb—nothing is as impossible to restore than a dried-out piece of white turkey meat!

4

Beef

BEEF HAS BEEN a favorite since biblical times. For centuries, cows, bulls, steers, and oxen have been carefully bred, nurtured, fed, and finally slaughtered to feed the populace in many parts of the world.

✐ TYPES OF CATTLE

There are many breeds of cattle as there are many grades in quality.

The breeds we prefer in our shop are Aberdeen-Angus (sometimes called Black Angus) and Hereford (frequently referred to as White Faces).

Aberdeen-Angus have no horns. Their bodies are bulky and black. Their legs are short and they have stumpy necks. This breed was first developed in the Highlands of Scotland near the North Sea. In 1873 an American cattleman imported some to his Kansas farm. Today the "Black Angus" is raised in the East and South as well as in the Middle West.

Hereford cattle have red bodies and white faces. They also have white patches on their chests, flanks, lower legs, and on the tips of their tails. Herefords have short necks and broad heads, and their legs are short and straight. They can be raised on grasslands in the West, but they become tastier if they are shipped to the Middle West when young and are fattened on corn. We especially like to buy Herefords as "baby beef" (eight to twelve months old and weighing from 700 to 1,100 pounds).

⁊ PRIME BEEF

Prime beef comes from steers that weigh from 900 to 1,300 pounds and range in age from one to two years. As far as we are concerned, the younger the steer the better. A thousand-pound steer yields approximately 465 pounds of net retail cuts, and only 75 pounds are steaks that are suitable for broiling.

⁊ THE AGING OF BEEF

Although fresh beef usually reaches retail stores from six to ten days after slaughtering, we do not sell it immediately. We age our prime beef by holding it in our coolers (34° to 38° F.) for four to six weeks. Such aging develops additional tenderness and excellent flavor. We use a rotating system. As soon as a side of beef has aged to what we consider perfection, we take it out of its cooler for butchering. Then we buy another steer from the wholesaler to replace it.

⁊ HOW TO BE A "PRO" IN BUYING BEEF

You can't go wrong in beef-buying if you know that your butcher handles only prime quality. But should you go to a new butcher, ask him about the grade of beef he carries, then take a good look at the meat before you buy it. There is a wide variation in the appearance of beef cuts. The quality and structure of meat in a butcher shop will give you an advance tip about the ultimate tenderness and flavor of the beef you will serve on your dining table.

COLOR AND TEXTURE

The best-quality beef has a minimum of outer fat, and

that is creamy in color. The bones are soft-looking and have a reddish coloration. The meat itself is firm, fine-textured, and is usually a light cherry red.

MARBLING (OR GRAINING)

Prime beef is marbled with a delicate interlacing of fat. The best marbling (or graining) is a fine-needle type that runs throughout the meat as if woven like a cobweb. It assures a high degree of tenderness. Lesser grades have a thick crayon-type graining. Thick marbling often results when steers are not slaughtered at an early age. Instead, they are overfed and fattened, which gives ranchers extra poundage (and profit) for wholesale selling. Remember that the thicker the marble the tougher the meat is apt to be.

Actually, the marble is a lubricant and dissolves into the meat as it is cooked. The silky, tiny threads in top prime beef are so delicately interlaced that they dissolve evenly, making the meat juicy and tender. When the marble is thicker and is uneven in distribution, the fat does not dissolve (or cook) as fast as the meat—making for toughness and fattier-tasting meat. This is true even when meat is boiled.

⸭ WHAT YOU SHOULD AVOID IN BUYING QUALITY BEEF

- ⸭ Fat that has a yellowish or grayish color
- ⸭ Beef with heavy marbling
- ⸭ Beef with absolutely no marbling
- ⸭ Meat that has a deep red color
- ⸭ Meat that has a two-tone coloration
- ⸭ Beef with a coarse texture
- ⸭ Beef with excessive moisture
- ⸭ Meat that is too fresh

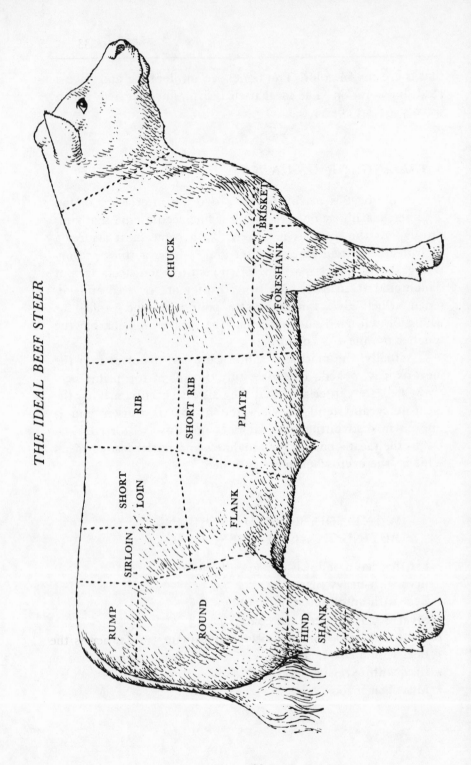

THE IDEAL BEEF STEER

CHUCK

BRISKET

FORESHANK

RIB

SHORT RIB

PLATE

SHORT LOIN

SIRLOIN

RUMP

FLANK

ROUND

HIND SHANK

BASIC CUTS OF BEEF

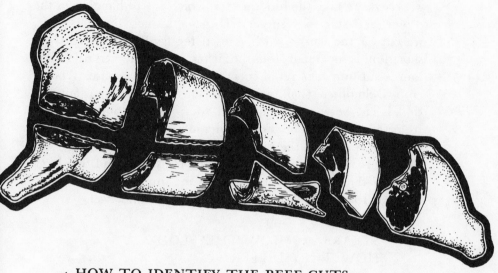

✦ HOW TO IDENTIFY THE BEEF CUTS YOU BUY

A beef steer is so large that it is impractical to transport it in one piece from packer to wholesaler to retailer. Therefore it is first split into sides. When we buy a side of prime beef, we hang it in an upright position on a V hook in one of our coolers. When it has aged to our satisfaction we separate the forequarter from the hindquarter. Then we place the forequarter on one of our butcher blocks and divide it into four sections.

FOREQUARTER

First we cut five bones from inside the chuck, which leaves a seven-bone rib roast. This is cut straight up and down. Then we cut the plate from the rib. Next we cut straight across the front of the forequarter, separating the chuck from the arm or shank. From these four basic cuts we later obtain: rib roasts, rib steaks, flanken, brisket, arm and shoulder steaks, as well as chuck (for steaks, pot roast, ground beef, and beef tea).

HINDQUARTER

Now we take the hindquarter, which is still hanging in the cooler, and start our surgery. Using a 12-inch butcher knife, we cut off the flank, which is used for flank steaks. Next the short loin is separated from the sirloin. Later the versatile short loin is cut into such retail cuts of steak as porterhouse, T-bone, club, Delmonico, shell, and fillet. Then a very important section is cut between the hip and rump. All sirloin cuts come from this portion. The final cutting job is to separate the round from the rump and the hind shank. The rump is used for long-cooking roasts; the hind shank is used for stews. And that large portion—the round—can be cut into any number of retail cuts: boneless roast beef, London broil steaks, ground beef.

⚹ STEAKS . . . HOW THEY LOOK . . . HOW THEY DIFFER

Steaks are probably the most popular meat item on the menu of any good restaurant. But there are so many types that it is well for retail buyers to know how they differ. To simplify your buying, we will give you the characteristics of each along with sketches and suggested cooking methods.

PORTERHOUSE STEAK

This is one of the most popular steak cuts, perhaps because it has a generous section of tenderloin. The porterhouse is cut from the short loin, nearest the sirloin. It is fine-grained, with a characteristic portion of fat, and is usually cut from 1¼ to 3 inches thick. Sometimes customers have the tenderloin removed so that they can serve it separately as a filet mignon. The tail can be removed and ground, then this "pattie" can be cooked separately, placed where the tenderloin has been removed, or tucked in between the bone and the fillet. In either case it should be tied in place so that it will not fall apart while cooking. See sketch, p. 25.

The porterhouse is usually broiled (in oven or outdoor grill), but the thinner cuts may be pan-broiled.

T-BONE STEAK

This steak is easily identified by its T-shape bone. It comes from the center section of the short loin, between the porterhouse and club. It is similar to the porterhouse but has a smaller section of tenderloin and a smaller tail, with a fine-grained shell. It should be cut from 1 to 3 inches thick. Broiling is the best cooking method, pan-broiling for thinner cuts.

CLUB STEAK

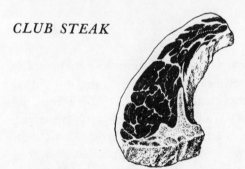

Sometimes called Delmonico, this steak is triangular in shape. It is smaller than the T-bone but has the same large "eye" section with no tenderloin. The club is cut from the short loin, next to the rib end. When cut properly, this is a delicious and tender steak. But beware: Butchers have been known to pass off rib steaks (which are lower in price) for clubs.

When you buy a club steak from a new butcher, take a good look at the steak's "eye." The meat should be fine in texture with delicate marbling. If the meat seems coarse and contains fat chunks, you will know it is not the quality you want. Another indication is that you will find heavier grain in the rib area. We find that our customers enjoy the club cut for minute steaks or our famous "beef scalp." For these types the steak is cut about an inch thick. But sometimes we cut a club steak about 2 or 3 inches thick and remove the bone. This makes an excellent steak for slicing diagonally after it has been broiled.

TENDERLOIN STEAK

The popular name for this steak is filet mignon. As mentioned previously, some customers like to have the tenderloin section removed from their porterhouse steaks, making individual fillets. But we sometimes have requests for a complete fillet strip.

FILLET STRIP

Fillet strips are removed from the short loin before any other steaks are cut. Sometimes we cut these into individual fillets in the thickness the customer wishes. These may be broiled or pan-broiled (see Tournedos of Beef with Béarnaise Sauce, p. 48). A complete fillet strip can be broiled and then cut into individual portions, or the whole strip can be covered with pastry and baked (see Beef Wellington, pp. 50–51).

The tenderloin is the most tender of all steaks. But no matter how delicious and tender, some find the texture too soft.

SHELL STEAK

When the tenderloin strip has been removed from the short loin, the remaining meat is then cut into individual steaks of the thickness desired. The correct name for such steaks is shell. However, restaurants are apt to call them by various

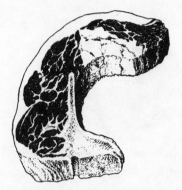

other names: New York strip, Kansas City strip, or just plain strip steak.

A shell steak is easily identified in your meat market. It looks exactly like a porterhouse or T-bone without the tenderloin. In other words, the eye of the meat is resting on the shell bone.

These steaks can be cut in any width you wish, from 1 to 2½ inches or thicker. Our customers frequently ask us to remove the bone. In this way the steak can be cut diagonally after broiling (see "Shell Roast," p. 42).

SIRLOIN STEAK

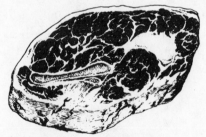

The sirloin is a large steak, which makes it suitable for large families or parties. The sketch shows a typical sirloin, usually cut from 2½ to 3½ inches thick, with a small amount of wedge-bone. Sirloins vary in shape and bone size. There are round-bone, pin-bone (or hipbone), and double-bone steaks. These can be cut from 1 to 3 inches thick. Any sirloin steak is tender, delicious, and excellent when broiled in the oven or barbecued. Of course, the thinner cuts may be pan-broiled.

SIRLOIN TIP STEAK

This cut is also referred to as a boneless sirloin. It comes from the bottom tip of the sirloin section. It is less tender than the sirloins with bones, but it has a delicious flavor. We usually cut it about 2 inches thick and suggest that it be braised for half an hour.

RIB STEAK

This is similar in appearance to club steak and is sometimes sold as such, even though it is less tender, has more fat, and should be less expensive. Rib steaks have an excellent flavor and, of course, come from the rib section. Since it is from the front portion of forequarter, this cut is often sold in kosher meat shops (for more about kosher meats, see p. 45).

CHUCK STEAK

Sometimes called blade chuck, this steak comes from the shoulder (or chuck) section of beef. It is a very economical cut with a well-developed flavor, but it varies in tenderness. We consider the first three bones of the chuck section the tenderest. They are adjacent to the rib roast and contain a sizable extention of the rib eye. These cuts are satisfactory for a family barbecue. The cuts that are farther down are less tender and should be cooked with moist heat, either braised or stewed.

FLANK STEAK

This is a lean, flat muscle with no bone at all. The meat fibers run lengthwise. There is only one flank steak to a side of beef. For best results, this steak should be broiled quickly and sliced on a slant. It has a lovely flavor and a tender texture when sliced in thin strips after cooking. This cut is mainly used for London broil, and comes from the lower section of the short loin.

ROUND STEAK

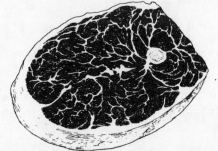

Of course, this cut comes from the rump (or round) section of the hindquarter. It is oval in shape and has a small, round bone, but we sell it without this bone. It has practically no fat, and because it is so lean it is excellent for steak tartar. Round steak makes a fine London broil (but this must be one of the first three cuts) and can also be used as a roast.

Since this cut of beef has very little waste, it makes an economical buy. However, because it lacks marbling, it is not as flavorful or juicy as other cuts.

GROUND BEEF

Should you go to a new butcher, please take our advice: Don't buy beef that has already been ground and is on display in a refrigerated cabinet. Such ground meat can come from any part of the animal—including trimmings! We recommend only four types of beef for grinding.

1. ROUND

As mentioned before, this has practically no fat. Aside from steak tartar, it makes a fine "diet burger"—because it is so lean. If you are dealing with a new butcher, the best way to get top quality is to use the advice of a mother who sent her young daughter shopping. She told her to ask for "a pound of the top of the round." After it had been cut and weighed, she told her to say, "Ground, please."

2. CHUCK

This is perfect for meatballs, meat loaf, and hamburgers. It is tasty and quite juicy because of its high fat content.

3. SIRLOIN

Even though this meat is well marbled, its tenderness and flavor make it sweet for raw eating (steak tartar) and de luxe for hamburgers.

4. TAIL OF PORTERHOUSE

Even though most meat markets do not make this type of

ground beef available to their customers, it has become one of our specialties. It is juicy and sweet-tasting and makes a truly delightful hamburger or meat loaf.

/ OTHER BEEF CUTS—ROASTS

PRIME RIBS OF BEEF

This is the most desirable and most tender of all beef roasts. These ribs, of course, come from the rib section of the forequarter. The meat is very juicy and is well marbled with a layer of fat on the outside. The size of this roast can be quite flexible. It can be cut to serve a small family (2 to 4) or a large dinner party (16). There are four ways we prepare prime ribs:

1. STANDING RIB ROAST

The roast is lightly trimmed and the short ribs are cracked.

2. HALF STANDING RIB ROAST NO. 1

This is also trimmed lightly, but the short ribs are completely removed.

3. HALF STANDING RIB ROAST NO. 2

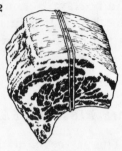

Again, the roast is lightly trimmed and the short ribs are taken off. All bones are removed from meat and then tied back in place. After roasting, strings are cut, bones are removed, and slicing is done with ease.

4. ROLLED RIB ROAST

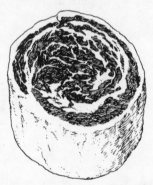

The rib roast is well-trimmed and short ribs are cut off. Then the roast is completely boned and rolled and tied over the outside layer of fat.

SHELL ROAST

When the tenderloin strip has been removed from the short loin, the remaining meat can be cut into individual shell steaks, or it can be cut into larger pieces to become roasts. These are very popular in our shop. They are excellent for parties and can be cut to the size needed for your guest list. This is a most desirable roast, especially when we bone it, making it very easy to slice.

SIRLOIN ROAST

Like sirloin steaks, this section of a steer is tender and delicious. When properly prepared by your butcher, it can be an excellent (and less expensive) alternate to a prime rib roast.

⁄ OTHER CUTS OF BEEF—POT ROASTS . . .
STEWING . . . BRAISING

SHOULDER ROAST

This is located at the bottom of the chuck and is normally sold in two sections. The front section has a round bone and is sometimes called arm roast. The back section is boneless. Although the front cut may be used as a roast beef (when price is a factor), both cuts are better for potting. The meat in the shoulder is usually lean and dry.

RUMP ROAST

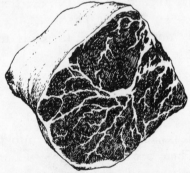

This roast is cut from the top end of the hindquarter. It is triangular when the bone is left in and usually rolled when the bone is removed. This cut has a moderate amount of fat and is tender. Even though some people use it for oven-roasting, we recommend it for pot roasts.

CHUCK ROAST

Another type of beef for pot roasts is chuck. It is usually sold with the bone left in but can be boneless. It has some fat, is juicy and well flavored, and, of course, comes from the fore-quarter of beef.

BRISKET

As shown in the diagram on p. 32, the brisket is in front of the foreshank and under the chuck. It is frequently cured for corned beef but is excellent for pot roast. There are two cuts of brisket. The first cut is a bit dry but lean, the second is much fatter and therefore more juicy.

PLATE

Even though the plate is comprised of layers of fat and is lean, this cut tends to be quite stringy. It can be sold flat or can be rolled. Because of its texture, it should be simmered in water very slowly until it is tender.

SHORT RIBS

We call this cut by its German name, flanken. These are cut from the ends of the rib roast and the plate. They contain layers of lean and fat with the flat rib bone. They make an exquisite boiled beef (pp. 83–84), enhance the flavor of soups, and are excellent for pot roast.

⸰ KOSHER MEATS AND POULTRY

Meat and fowl are made kosher by the manner in which the animals and fowl are slaughtered. The most important factor is the removal of the jugular vein, which permits the blood to drain off. Kosher meat is indicated by the symbol "K" as well as by the USDA inspection stamp. The slaughtering of the animals and fowl must be done under the supervision of a rabbi.

Such meat and fowl is sold primarily in kosher butcher stores; however, many supermarket chains sell both kosher and non-kosher products. Orthodox Jews, who abide by strict dietary laws, will trade only in stores where kosher meat is sold exclusively. Basically, the word "kosher" means "clean." All kosher meat is supposed to be consumed within seventy-two hours after slaughtering. After that period it must be washed in salted water—to recover the kosher quality.

CLASSIC BROILED PORTERHOUSE STEAK

Serves: 4
Time: 12 to 14 minutes on each side for rare

1 2-inch-thick porterhouse steak
¼ lemon
1 tablespoon oil
salt and freshly ground pepper

2 cloves garlic (crushed)
½ teaspoon scallions (finely chopped)

Be sure to have butcher score the outer fat. Rub fat with lemon (this will prevent burning or smoking). Mix oil, salt, pepper, garlic, and scallions and smooth over steak on both sides. Allow to marinate for about 2 hours at room temperature. Preheat broiler. Place in broiler and cook according to doneness you wish.

✦ *Serve with baked potatoes, stewed tomatoes, and a husky dry red wine.*

BROILED FILLET STRIPS WITH FANTASTIC SAUCE

Mr. and Mrs. Douglas Auchincloss delight in good food, so their fine chef caters to their good taste. Here is a dish they particularly admire. It is plainly and simply broiled strips of beef fillet, but the sauce makes all the difference.

Serves: 4
Time: About 15 minutes

2 pounds fillet of beef
½ cup chili sauce
½ cup ketchup
dash paprika
dash cayenne pepper
pinch basil
pinch tarragon
pinch chervil
pinch oregano

pinch sage
2 tablespoons chives (chopped)
1 teaspoon Worcestershire sauce
2 teaspoons soy sauce
salt and freshly ground pepper
3 tablespoons dry red wine
3 tablespoons Grand Marnier
3 tablespoons Drambuie

Have butcher cut fillet into strips and bring to room temperature. Preheat broiler. Meanwhile, make the sauce. Blend the chili sauce with the rest of the ingredients (except Grand Marnier and Drambuie). Pour this mixture into a sauce pan and place over low heat, stirring constantly. While this is simmering, place fillet strips under broiler heat and cook (depending on thickness) for a few minutes on each side. Now add the Grand Marnier and the Drambuie to the sauce. Place fillet strips on a hot platter and pour the sauce over them.

⸱ *Although the Auchinclosses have not told us what they would like with this dish, we are sure they would approve of rice or noodles with that delicious sauce.*

BUTTERFLY FILET MIGNON WITH SHERRY

Serves: 4
Time: About 25 minutes

4 fillets (8 ounces each)	⅛ pound butter
6 tablespoons oil	salt and freshly ground pepper
1 large pimiento (thinly sliced)	4 tablespoons sherry wine
1 green pepper (thinly sliced)	

Ask butcher to "butterfly" the fillets so that they are about ¼ inch thick. Bring them to room temperature. Meanwhile, heat 4 tablespoons of the oil in fry pan and add pimiento and pepper; cover and let simmer for about 15 minutes. In another large fry pan heat butter and rest of oil; add salt and pepper. When bubbling furiously, add fillets and brown on each side (about 4 minutes). Turn down heat and add ingredients from other pan and the sherry wine. Cover and simmer for about 5 minutes.

⸱ *Serve with asparagus with lemon butter and claret wine.*

TOURNEDOS OF BEEF WITH
BÉARNAISE SAUCE

Serves: 6
Time: About 8 minutes (for meat)

6 fillets
6 strips bacon
2 tablespoons butter

salt and freshly ground pepper
6 rounds of bread (toasted)
Béarnaise Sauce (p. 284)

Have butcher cut beef tenderloin into 6 fillets 1½ inches thick. Cook bacon strips until partially done. Drain on paper towels and discard bacon fat. Wrap a strip of bacon around each fillet and use skewer to hold in place. Melt butter in frying pan. When bubbling, sauté the fillets for 3 minutes on each side (for rare). Place each fillet on a slice of crisp toast and cover with Béarnaise Sauce.

/ Serve with broiled tomatoes, watercress salad, and Burgundy wine.

FILET MIGNON WITH PEPPERCORNS
AND BRANDY

Duke Fulco Di Verdura is an Italian and loves dishes from his native Sicily, but he is a gourmet with international tastes. He particularly likes this filet mignon.

Serves: 4
Time: About 5 minutes

4 thick slices of beef fillet
salt
peppercorns (about 16)
4 tablespoons butter

1 clove garlic (minced)
4 slices bread
3 tablespoons brandy

Dry fillets on paper towels and sprinkle with salt. Punch tiny holes in meat and insert peppercorns on each side. Place 1 tablespoon of butter in a heavy frying pan. When sizzling, add the garlic. When transparent, add the fillets and brown well on both sides. Meanwhile, in another frying pan, melt 2 tablespoons of butter and sauté the bread. When brown on both sides, remove to paper towels and then to a hot platter. By now the meat should be browned. Add the last tablespoon of butter. When it has melted, add the brandy and set aflame. Cook for half-minute after flame has died down. Place fillets on fried toast and sprinkle pan sauce over all.

⌐ *Serve with a hearty mixed green and vegetable salad and imported red wine.*

FILET MIGNON WITH COGNAC AND HERB SAUCE

Serves: 2
Time: About 8 minutes

Filet

2 2-inch fillets	2 bread slices
2 tablespoons cognac (warmed)	

Herb Sauce

¼ pound butter (softened)	1 teaspoon parsley (chopped)
1 clove garlic (crushed)	salt and freshly ground pepper
¼ teaspoon dill	

Preheat broiler. The herb sauce should be made while the fillets are resting at room temperature (for at least half an hour). Blend the softened butter with the garlic, dill, parsley, and salt and pepper. Place the 2 thick fillets in a shallow pan that will fit under the broiler. Then pour the warmed cognac over both sides of the meat and light it. When the flame goes out, spread half of the herb sauce over the top side of the meat, which is then placed under the broiler for 3 minutes. Turn it and scoop up some of the butter on top and broil for another 3 minutes (or more, if you want it more well-done). Meanwhile, toast the bread after you have cut it into the sizes of the fillets. Spread them, while still hot, with the rest of the herb butter. Then place the fillets on top, adding a little juice from the pan.

⅟ *This deserves exquisite accompaniments. How about artichoke hearts (hot or in a salad) and champagne?*

BEEF WELLINGTON

The "21 Club" is a restaurant that has been highly rated for many years by gourmets far and wide. Old-timers still refer to it as "Jack and Charlie's" because the originators were Jack Kriendler and Charlie Berns. And to anyone who frequents this noted establishment the names of Berns and Kriendler are by-words. Younger brothers of these two still carry on. It's now Jerry and Bob— H. Jerome Berns and I. Robert Kriendler. The latter is a customer of ours and he gave us this excellent recipe for Beef Wellington.

Serves: 6 to 8
Time: About 1 hour

1 4-pound fillet of beef	2 egg yolks
oil	¾ cup water
salt and freshly ground pepper	pâté de foie gras
¾ pound butter	truffles (diced)
1 pound flour	1 egg yolk
½ ounce salt	¼ cup milk
1 ounce sugar	

Preheat oven (450° F.). Dry the fillet on paper towels and rub all sides lightly with oil. Season with salt and pepper. Place in open baking pan and bake for 40 minutes. Meanwhile, chop the butter into the flour along with the salt and sugar. Mix the egg yolks with water and add gradually to dry mixture. Gather dough together and roll it to a size that will encompass the fillet. When the beef is ready, remove from pan and cover with foie gras and a few diced truffles. Place beef on top of dough and roll the dough around it so that it is completely surrounded.

Now combine 1 egg yolk with milk and brush on top of the dough. For an attractive look, dough may be lightly scored in a crisscross pattern before adding the egg-milk brushing. Place in oven for 20 minutes or until the dough has become a golden brown.

⟋ *Serve with endive salad and a superb dry red wine.*

BAKED FILLET OF BEEF

Serves: 8
Time: 35 to 40 minutes

4 to 4½ pounds fillet
1 clove of garlic (cut in half)

2 tablespoons chicken fat (softened)
salt and freshly ground pepper

Preheat oven (450° F.). In the ends of fillet, place half a clove of garlic. Then spread chicken fat on all sides and sprinkle with salt and pepper. Place fillet in uncovered pan and bake for half an hour. Test with small sharp knife for rareness and cook longer if necessary.

⟋ *Serve with small paprika potatoes (these can be cooked in oven with fillet), Bordelaise Sauce No. 2 (p. 285), and a delicate claret wine.*

EASY LONDON BROIL

Serves: 4
Time: 6 minutes each side for rare

1 1-inch top-of-round steak
2 tablespoons olive (or other) oil

1 garlic clove (crushed)
1 teaspoon oregano
salt and freshly ground pepper

Ask butcher to give you the first cut of top-of-round. Place meat on waxed paper. Meanwhile, mix rest of ingredients thoroughly. Spread this mixture on both sides of steak and let stand for about 2 hours at room temperature. Turn once or twice so that marinade can soak in. Preheat broiler. Broil for the doneness you prefer. Transfer steak to hot platter and cut diagonally.

⨍ *Serve with stuffed baked potatoes with sour cream and chives, a mixed green salad, and a hearty red wine.*

GRILLED FLANK STEAK

Serves: 4
Time: About 15 minutes

2½ pounds flank steak
1 tablespoon oil
1 shallot (finely chopped)

salt and freshly ground pepper
1 clove garlic (crushed)

Be sure to ask the butcher to score the steak crosswise. Mix the oil, shallot, salt and pepper, and garlic together. Rub this mixture over both sides of the steak and let stand at room temperature for about 2 hours. Preheat broiler. Broil 5 minutes on each side for rare, more for medium and well-done.

⨍ *Tip: Slice on the bias, about ¼ inch in thickness.*
⨍ *Serve with sautéed mushrooms, a delicate green vegetable such as spinach, and a light red wine.*

BEEF SCALLOPINI ROULADES

Serves: 6
Time: About 1½ hours

12 slices beef scallopini
Stuffing for Beef Scallopini
  Roulades (p. 281)
3 tablespoons butter
12 small white onions
⅓ cup flour

1 tablespoon tomato paste
10 ounces beef bouillon
2 cups Burgundy wine
1¼ pounds mushrooms
8 chestnuts (quartered)

Preheat oven (350° F.). Place beef scallopini on wax paper

and flatten. Spoon out the stuffing on each piece. Roll up meat so that stuffing does not ooze out and tie with string. Heat butter in a Dutch oven. When sizzling, add the roulades and brown on all sides (about 5 minutes). Remove meat and add onions and stir in flour and the tomato paste. When the sauce has blended well, gradually add the bouillon and wine. Stir constantly until sauce bubbles, then return beef to Dutch oven and add mushrooms and chestnuts. Cover and place in over for about 1½ hours. (And, of course, remove string before serving.)

ʃ *Serve with noodles, crisp green salad, and Burgundy.*

BEEF SCALLOPINI WITH PROSCIUTTO

Serves: 6
Time: About 40 minutes

6 thin slices beef scallopini	salt and finely ground pepper
6 thin slices prosciutto	3 tablespoons butter
½ cup cooked rice	1 tablespoon cognac
1 cup cooked mushrooms (chopped)	3 shallots (finely chopped)
1 scallion (finely chopped)	6 sage leaves (or 1 teaspoon powdered)

Flatten scallopini on wax paper and place a slice of prosciutto on top of each. Meanwhile, mix rice, mushrooms, scallion, and salt and pepper. Spoon this mixture on top of prosciutto. Now roll up each slice and tie firmly with string. Place rolls in sizzling butter and brown on all sides (about 10 minutes). Meanwhile, combine cognac, shallots, and sage. Reduce heat and add to pan, mixing well with butter. Simmer for 15 minutes, then turn rolls and simmer for another 15 minutes. Add more butter if pan has dried out.

ʃ *Serve with buttered carrots and string beans (cooked in milk and flavored with rum) and Burgundy wine.*

STUFFED STEAK ROLLS

Serves: 4 to 6
Time: About 35 minutes

2 pounds boneless sirloin
½ pound pork sausage
  (cooked)
½ cup breadcrumbs
1 egg
1 tablespoon capers

2 cloves garlic (crushed)
salt and freshly ground pepper
3 tablespoons oil
½ cup water
1 tablespoon marjoram
1 medium onion (sliced)

Have butcher cut steak into paper-thin slices. Flatten steak slices on wax paper. Meanwhile, remove skin from sausages and break up into small pieces. Mix with breadcrumbs, egg, capers, garlic, and salt and pepper. Spread this mixture on steak pieces, then roll up and fasten with toothpicks. Heat oil in heavy fry pan and add steak rolls. Sauté until brown on all sides (3 to 5 minutes). Drain off and add water, marjoram, and onion slices. Cover and simmer for 30 minutes, adding water or wine if pan dries out.

⸙ *Serve with Brussels sprouts with lemon butter and a Burgundy wine.*

KOREAN BULGOGI (SIRLOIN STRIPS)

Serves: 6
Time: 20 minutes

3 pounds sirloin
2 tablespoons sesame seeds
4 ounces soy sauce
4 scallions (finely chopped)

2 cloves garlic (crushed)
1 tablespoon sugar
salt and freshly ground pepper
4 ounces water

Have butcher cut sirloin in paper-thin strips. Brown sesame seeds in large fry pan. Remove from heat and add other ingredients (except meat) and stir well. Then add meat and marinate for 2 hours or more. Place pan over medium heat and cook for 20 minutes. If sauce seems too watery, dissolve a tablespoon of arrowroot in a little water or wine and add.

⟩ *Serve with mounds of rice and a green salad. A medium-sweet red wine.*

BEEF ORIENTALE

Dr. Nathaniel Troy Kwit, one of America's leading heart specialists, is an epicure. Food is his hobby and he is a national officer of that elegant dining society, Chaine des Rotisseurs. Here is a recipe he enjoys cooking.

Serves: 6
Time: About 35 minutes

2 pounds beef fillet	2 tablespoons oil
salt and freshly ground pepper	1 tablespoon butter
3 tablespoons oil	24 mushroom caps
2 medium onions (chopped)	6 small tomatoes (quartered)
2 cloves garlic (minced)	3 tablespoons cornstarch
3 green peppers (thinly sliced)	1½ tablespoons soy sauce
2 cups beef bouillon	½ cup water

Cut meat into narrow strips about 2 inches long and ½ inch thick (porterhouse may be used instead of fillet). Season meat with salt and pepper. Heat 3 tablespoons of oil in a large skillet. Add the meat and onions and cook quickly until meat is brown on all sides. Add the garlic just before the meat is finished browning. Add the green peppers and bouillon. Cover pan, reduce heat, and simmer for 10 minutes. Meanwhile, heat the 2 tablespoons of oil and butter in another large skillet until sizzling. Add the mushrooms and sauté until tender. Add the meat mixture and then the tomatoes. Reduce heat and simmer for 2 minutes. Mix the cornstarch, soy sauce, and water and stir gradually into the meat mixture. Cook, stirring until the mixture has thickened.

⟩ *Serve with hot rice and Beaujolais wine, chilled very slightly.*

STEAK STRIPS WITH VEGETABLES

Serves: 6
Time: 25 minutes

2 pounds boneless round steak
3 tablespoons oil
salt and freshly ground pepper
½ teaspoon ginger
1 clove garlic (crushed)
1 package frozen string beans
(thawed)

2 green peppers (cut in
eighths)
1½ cups chicken bouillon
3 tablespoons cornstarch
2 tablespoons soy sauce
2 tomatoes (cut into eighths)

Have butcher cut steak into thin strips. Heat oil in large fry pan or Dutch oven. Stir in salt and pepper, ginger, and garlic, then add meat strips. Sauté until brown on both sides (3 to 5 minutes). Reduce heat and add string beans, peppers, and bouillon. Simmer covered until beans are tender but still crisp (6 to 8 minutes). Meanwhile, mix cornstarch and soy sauce with a little cold water. Add this mixture and tomatoes and stir constantly until sauce has thickened (about 5 minutes).

⁊ *Serve with mashed potatoes, green salad, and a dry red wine.*

STEAK KEBAB WITH CREAM SAUCE

Serves: 4
Time: About 20 minutes

3 pounds sirloin (cut into
28 1-inch squares)
28 thin slices Canadian bacon
4 tablespoons oil
2 tablespoons brandy
½ cup cream
4 tablespoons beef broth

salt and freshly ground pepper
1 teaspoon chopped dried
apricot
2 teaspoons butter (softened)
2 teaspoons parsley (finely
chopped)

Bring steak to room temperature. Wrap each square with a piece of Canadian bacon and secure with toothpick. Thread 7 of these on each of 4 serving skewers. Heat oil in large fry pan, then add skewers of meat and brown lightly on all sides (about 5 minutes). Reduce heat. Sprinkle brandy on meat and add the cream and broth. Then sprinkle salt and pepper and apricot. Simmer for 10 to 15 minutes, but be sure heat is low enough so that cream does not curdle. Now remove skewers and place on hot serving platter. Add butter to sauce and stir constantly until well combined (about 4 minutes). Pour sauce over steak and decorate with chopped parsley.

⁊ *Serve with fluffy rice, green peas, and rosé wine.*

FILLET TIDBITS WITH YAMS

Serves: 4
Time: About 1 hour

4 yams (large)
¼ pound butter (soft)
¾ pound fillet (cut in ½-inch cubes)

salt and freshly ground pepper
4 tablespoons Mushroom Sauce for Meat (p. 289)
½ teaspoon chopped parsley

Preheat oven (350° F.). Bake yams in oven for 45 minutes. While they are cooking, melt ⅛ pound of butter in a skillet and brown the fillet pieces on all sides and let them rest away from heat. After removing yams from oven, turn up heat to 450° F. Cut a good slice off the tops of yams and scoop out the pulp—when they are cool enough to handle. Mash this pulp with the other ⅛ pound of butter and season with salt and pepper. Place this mixture back into the yam shells, making an indentation in the center with a soup spoon. In this hollow space, place the fillet tidbits and add a tablespoon of mushroom sauce on each. Sprinkle with chopped parsley and place back in oven for about 10 minutes.

⁊ *Tip: Idaho potatoes can be used in place of yams.*
⁊ *Serve with a light green vegetable or salad, or even a spiced fruit. And a dry red wine, or even beer, would be tasty.*

BEEF STROGANOFF

Serves: 6
Time: About 15 minutes

2 pounds beef tenderloin
  (thinly sliced)
4 tablespoons butter
1 medium onion (chopped)
½ pound mushrooms (sliced)

salt and freshly ground pepper
2 cups sour cream
2 teaspoons prepared mustard
paprika

Cut tenderloin slices into strips. Place butter in large skillet and, when bubbling, add the meat and brown quickly on each side. Remove meat and keep warm and add the onions. When these are golden, add the mushroom slices and stir until they are wilted. Return the meat and season with salt and pepper and reduce heat. Add half of the sour cream and mix mustard with the other half. Add this to the mixture and stir until it is quite hot, but do not bring to a boil. Garnish with paprika.

⸱ *Serve with rice, buttered asparagus, and a chilled rosé wine.*

STEAK AND KIDNEY STEW

Serves: 8
Time: About 2 hours

2 veal kidneys
4 pounds sirloin
1 can beer
6 tablespoons butter
1 pound mushrooms (sliced)

2 onions (finely chopped)
8 slices prosciutto
salt and freshly ground pepper
2 level tablespoons flour
1 cup Marsala wine

Fat and membranes should be removed from kidneys, then they should be cut into ½-inch pieces. Sirloin should be cut into ¾-inch squares. Place kidneys and beer in bowl and let stand for about an hour. Now rinse them in cool water and dry well on paper towels. Heat 2 tablespoons of the butter in a fry pan; add mushrooms, cover, and allow to simmer. In another large fry pan, heat the other 4 tablespoons of butter. When sizzling, add the dried kidneys and sirloin and brown on all sides. Push to side of pan and add the onions. When they are golden brown, add the mushrooms and prosciutto and season with salt and pepper. Stir well, reduce heat, cover pan, and simmer for about 1½ hours, stirring now and then. Cool ½ cup of pan juice and mix with flour. Add this and wine to mixture in pan. Cover again and simmer for 15 to 20 minutes. If sauce is not juicy enough, add a bit of wine or water.

⁊ *Pour stew over crisply toasted English muffins, then have a green salad and dry red wine.*

STEAK AND KIDNEY PIE

Serves: 6
Time: About 2½ hours

3 pounds sirloin
1 pound veal kidneys
4 tablespoons flour
salt and freshly ground pepper
4 tablespoons butter
2 medium-size onions (chopped)
½ green pepper (chopped)
1 stalk celery (chopped)

½ pound mushrooms (chopped)
1 cup beef bouillon
½ cup port wine
dash of cayenne pepper
1 tablespoon pistachio nuts (crushed)
pie crust

Have butcher cut sirloin into 1-inch cubes. All fat should be removed from kidneys, then they should be refrigerated and allowed to soak overnight in slightly salted water. The next day, dry kidneys, remove membranes, and cut into ½-inch cubes. Mix flour with salt and pepper and dredge all meat well. Heat 2 tablespoons of the butter in a large fry pan and brown meat on all sides. Meanwhile, heat rest of butter in another fry pan and sauté the onions, pepper, and celery until onion is golden, then add the mushrooms. Reduce heat and simmer until mushrooms are limp, then combine vegetables with meat and add bouillon, wine, and cayenne. Stir until concoction is bubbling, then reduce heat and simmer for about 2 hours. Stir from time to time. Should the juice dry out, you may wish to add a bit more wine and reduce the heat somewhat. At the end of cooking, if you find that the sauce is a bit too watery, you can easily adjust this by cooling some of the sauce and then mixing a teaspoon or so of arrowroot in it. When the sauce is just right—a medium thickness—place the stew into a deep baking dish and allow it to cool. Sprinkle the pistachio nuts on top. Preheat oven (450° F.). Meanwhile, roll out enough pie crust to cover baking dish. Pinch the edges of the pie dough around edges of dish and then make a few vents on top of pastry to allow steam to escape. Bake for about 15 minutes, then lower heat to 350° F. and continue baking until crust is golden brown.

⸶ *This is such a substantial dish that it does not need starches or vegetables, but pears saturated with port wine would be a good dessert.*

BRAISED SWISS STEAK

Serves: 4
Time: About 1¼ hours

1½ pounds round steak	2 tablespoons oil
(¾ inches thick)	8-ounce can tomato soup
¼ cup flour	½ cup water
salt and freshly ground pepper	

Dry steak on paper towels. Mix flour with salt and pepper and pound into steak with meat hammer or edge of heavy saucer. Heat oil in large skillet and brown steak on both sides (about 10 minutes). Lower heat and add soup and water. Cover and allow to simmer for about 1 hour or until meat is tender. Stir from time to time and add more water if sauce dries out.

⚹ *Serve with creamy mashed potatoes, string-bean salad, and dry red wine.*

STEAK IN THE BLANKET

Yield: 16 portions
Time: 10 to 13 minutes

1½ pounds sirloin steak (cut in 16 finger-size pieces)

1 cup garlic dressing (see p. 290)
8 refrigerated crescent-type rolls

Marinate steak strips in garlic dressing for several hours at room temperature. Drain well on paper towels, then brown quickly on all sides in hot fry pan. Remove from heat. Preheat oven (350° F.). Meanwhile, unwind crescent rolls on wax paper —each roll will become a large triangle. Cut each of these into 2 smaller triangles. Flatten the dough with your hands and place a steak finger at wide end of each triangle. Then wind so that each becomes a plump little crescent roll. Place on cookie sheet and bake about 10 minutes or until golden brown. Turn off oven and cover rolls with aluminum foil until ready to serve.

BEEF FONDUE BOURGUIGNONNE

The word "fondue" is derived from the French word *fondre*— meaning to melt. The Random House Dictionary defines "fondue" in this way: "a saucelike dish of Swiss origin, made with melted cheese and seasonings, together with dry white wine, and sometimes eggs and butter, usually flavored with kirsch and served as a hot dip for pieces of bread." But the idea of dipping food into a central pot probably started long before the Swiss dreamed of it. The

Chinese have known about the advantages of a communal pot for centuries. It was particularly popular in the cold-weather months, when the guests would hover around a pot kept bubbling over a charcoal fire. Into this they would dip pieces of food with their chopsticks.

Today the fondue pot has become the rage the chafing dish was a few years ago. It has the same heating arrangement as that of the chafing dish—which can be fired by a canned heat cooking fuel— but the shape of the pot is different in that the sides of the top slope inward to prevent splattering.

Long forks with insulated handles are used to cook individual portions of meat. Chopsticks are also excellent, if your guests are proficient in handling them. Should you use chopsticks, be sure to first soak them in water so they will not burn in the hot oil. Once you have served a fondue, you will realize that it is better to have an extra pot if there are more than four persons. It becomes rather confusing if there are more than four forks in the pot at the same time.

TABLE SETTING: The long forks or chopsticks are used only for cooking; therefore each guest should have a regular fork, as well as a plate. There are individual plates that have slight divisions so that the various sauces can be spooned out and ready for dipping when each person's meat is cooked. Should you not have a "lazy-susan" with its own sauce dishes, you will need enough small bowls for the number of sauces you wish to serve. For a more casual party, each guest may wish to dip his meat in any of the central bowls he prefers instead of spooning a this-or-that onto his plate.

INGREDIENTS: Even though this recipe is called Beef Fondue, many other meats can be used to vary this delightful meal. And even if you decide to use some scallops or shrimp or hunks of cheese along with the meat, they may all be cooked in the same oil without affecting its flavor. Every item you decide to use should have all fat or sinews removed and then cut into bite-size cubes (¾ to 1 inch). Before the party, each cube should be dried well (moisture causes splattering) and brought to room temperature (about 1 hour). Here are some of our favorites:

fillet
sirloin
shell
calf's liver
chicken livers
chicken (no skin)

lamb (eye of loin or rib)
sausage (uncooked)
meatballs (cheese in center)
veal kidney
chunks of cheese

If you are planning a variety of meats, place each in a separate bowl and figure about 6 to 8 ounces of meat for each person. For bite-size pieces, allow 15 to 45 seconds of cooking time for each. But everyone is his own cook—and he or she can figure the time element from experience.

WHAT TO DO: Preheat oil on top of stove until it is 425° F. If you do not have a thermometer, throw a small piece of bread into the oil. When it browns in 1 minute, you will know that it is ready. Be sure your fondue pot is ½ to ⅔ full. Now place pot over cooking fuel. Of course, by this time your table is filled with all the various meats and the sauces you like best. And you will wish to have some attractive garnishes.

Sauces (pp. 284 and 293–296)

Horseradish and Bacon
Curry
Roquefort and Almond
Pickle and Caper

Hot Mustard and Olive
Garlic and Mushroom
Béarnaise

Garnishes

onions (minced)
parsley (chopped)
hard-cooked eggs (chopped)
almonds (slivered)
green pepper (chopped)
pimiento (chopped)

chives (chopped)
olives (chopped)
carrots (chopped)
coconut (shredded)
breadcrumbs

✓ *Tip: Should the oil smoke a bit, throw in a few cubes of bread from time to time.*
✓ *Serve with a large green salad with simple dressing (there are many spicy flavors in the sauces), hot and crisp Italian bread, and chilled beer.*

STEAK SUPREME

Emily Wilkens has received many awards for her accomplishments in the world of fashion. Her interest today is in youthful beauty and health. Her book, *A New You* (Putnam), is a bible of beauty and grooming. In her syndicated newspaper column, "Beauty and Nutrition," Miss Wilkens says, "The food you eat is the greatest source of beauty revenue you can acquire." She is an advocate of what she calls "live" foods—meat and vegetables cooked in a minimal way, which, she is certain, will make people feel "more alive." "Overcooking," she states, "destroys the important vitamins and enzymes." In her Steak Supreme the chopped steak is completely uncooked.

Serves: 4
Time: About 1 hour

2 pounds top-of-round
4 large Idaho potatoes
unsaturated oil (corn or sesame)
4 tablespoons fresh caviar
2 tablespoons chives (chopped)

Ask the butcher to chop meat very finely. Preheat oven (350° F.). Scrub potatoes and dry well, then coat skins with oil. Place potatoes in oven and bake for 45 minutes. Then take them out and, when cool enough to handle, cut them in half, scoop them, and rub insides of shell with oil. Return the shells to oven for 10 to 15 minutes. When they have crisped up, remove them from oven and, if they are too dry, rub a little more oil on their insides. Cool them and then fill each shell with chopped meat. Garnish each with caviar and chives.

✓ *Serve with raw-spinach salad with grated raw beets. The dressing for this should be quite light, just lemon juice and safflower oil. Also include a light chilled rosé wine (Alinca is her favorite) or beer in wine glasses.*

BEEF BLINTZES

Yield: 10
Time: About 30 minutes

Meat Filling

1½ pounds chuck steak
  (ground)
1 tablespoon chicken fat
2 tablespoons onion (chopped)

2 tablespoons celery (chopped)
salt and freshly ground pepper
1 egg (beaten)
½ cup bouillon

Dough

3 eggs (beaten)
1½ cups potato starch
½ teaspoon salt

¾ cup orange juice
1 tablespoon chicken fat

Bring meat to room temperature. Heat chicken fat in fry pan until bubbling. Sauté onions and celery until golden brown and remove from heat. Add salt and pepper to egg and stir into bouillon. Now add this to steak and add onions and celery when cooled. Let stand (so meat can absorb flavorings) while you mix dough.

Into well-beaten eggs, blend potato starch and salt and mix well. Gradually add the orange juice and stir until it becomes a thin batter. Use a little of the chicken fat to grease an 8-inch fry pan. Place over medium heat and pour in ¼ cup batter. Toss pan so that batter spreads evenly over bottom. When lightly browned (lift up a corner to make sure), place, cooked side up, on a paper towel. Immediately place 2 heaping tablespoons of meat filling on top of cooked side, then roll up and tuck ends in. Repeat this procedure until all batter and meat is used up.

Melt chicken fat in a very large fry pan. When sizzling, add the blintzes and brown all sides. If necessary, add more chicken fat.

⁊ *Tip: If blintzes are not to be served immediately, drain them on paper towels and place on a cookie pan. Cover them with aluminum foil and put them into a very low-heat oven to keep warm.*

SWEDISH MEATBALLS

Serves: 4
Time: 15 to 20 minutes

1½ pounds chuck steak
  (ground)
2 eggs (beaten)
¼ teaspoon thyme
¼ teaspoon basil
salt and freshly ground pepper
2 cloves garlic (minced)

3 anchovies (chopped)
1 small onion (grated)
½ cup Italian seasoned
  breadcrumbs
½ teaspoon chopped parsley
3 tablespoons water
1 can beef bouillon

Place meat in a large bowl and break up lightly with 2 forks. Add eggs and stir. Now add all other ingredients except bouillon. Mix together well and let stand at room temperature for about an hour so that meat has absorbed all of the flavorings. Dampen hands with oil or cold water and shape into balls the size of a walnut. Meanwhile, heat bouillon in large pan. When simmering, add meatballs and cook for about 10 minutes.

⸫ *Tip: These may also be served as canapés. In that case keep them warm in a very low oven. Then, just before serving, drain on paper towels, sprinkle with chopped parsley and pierce with cocktail picks.*
⸫ *Serve with creamed spinach, broiled tomatoes, and a rosé wine.*

STUFFED MEATBALLS—BAKED

Serves: 6
Time: About 20 minutes

1½ pounds sirloin (finely
  ground)
salt and freshly ground pepper
6 strips of bacon
6 small white onions (diced)

6 green olives (chopped)
3 slices cooked ham (diced)
1 tablespoon butter (soft)
1 cup breadcrumbs

Preheat oven (450° F.). Sprinkle ground sirloin with salt and pepper and mix with a delicate hand. Now divide the meat into sections and form round balls (about 2 inches round) with your hands (which have been covered with water or oil). In the center of each ball, make an indentation with your index finger (this hole should go down about ¾ of depth of ball). Meanwhile, sauté the bacon until it is quite crisp. Remove it to paper towels and crumble. (The bacon fat can be discarded or saved for future use.) Mix the crumbled bacon, onions, olives, diced ham, and soft butter. Stuff this mixture into the cavity of the meatballs, then smooth some of the meat over the opening to seal it. Scatter the breadcrumbs evenly on a sheet of waxed paper, then roll the meatballs so that all sides are covered. Now place the balls on a flat pan and cook in oven for 5 minutes. Reduce heat to 350° F. and cook for another 15 minutes.

f Serve with your favorite potatoes, a green vegetable or salad, and a dry red wine.

HOT BEEF-BALL CANAPÉS

Serves: About 10 (30–40 balls)
Time: About 15 minutes

1½ pounds sirloin (ground)
½ cup milk
¾ cup breadcrumbs
¼ cup dry red wine
1 medium-size onion (chopped)

¼ teaspoon ground ginger
salt and freshly ground pepper
sprinkle of garlic powder
6 strips bacon

First place meat in large bowl and let it come to room temperature. In a separate bowl, mix milk with breadcrumbs and add to meat along with all of the other ingredients (except bacon). Mix lightly with 2 forks and then form into walnut-size balls. The best way to do this is to put water or oil on your hands. Meanwhile, cook bacon on medium fire. When golden (but not brown), remove and place on paper towels. Now sauté the meatballs on all sides in the bacon fat. Then place them on paper towels to drain before putting them into a pan in the lowest possible oven temperature to keep warm until party time. Just before serving, crumble bacon and sprinkle over beef balls.

ORIENTAL CHOPPED MEAT
WITH VEGETABLES

Serves: 6
Time: About 30 minutes

1½ pounds sirloin (chopped)	1 cube beef bouillon
2 cloves garlic	½ cup boiling water
3 tablespoons oil	1 egg (slightly beaten)
2 carrots (diced)	2 tablespoons flour
1 large onion (diced)	2 tablespoons soy sauce
4 celery stalks (diced)	2 tablespoons water
2 medium-size green peppers	½ teaspoon dry mustard
(sliced)	salt and freshly ground pepper

Bring chopped steak to room temperature. Place garlic in large fry pan sizzling with oil until lightly browned. Remove and add steak to oil. When lightly browned, add vegetables and allow to simmer. Meanwhile, dissolve bouillon cube in ½ cup of water, add beaten egg, and stir until "cooked." In a small bowl, blend flour, soy sauce, water, and mustard. Combine these 2 mixtures and add to the meat and vegetables. Season with salt and pepper and stir well. Cover pan and continue simmering for about 20 minutes. Take a look at this mixture while it simmers. If it seems to have thickened too much, add a bit of water and a dash or two of soy sauce and stir well.

⸙ *Serve with flaky rice. No other vegetables are needed, but a crisp salad would do well, and so would a dry red wine.*

WEST INDIAN CHILI CON CARNE

Jawn A. Sandifer is a Justice of the Supreme Court of the State of New York and an excellent judge of fine cooking. Mrs. Sandifer is a talented cook, and even though she is very active in charity and committee work, each week she finds time to make a traditional "batch" of bread. Here is one of the couple's favorite recipes.

Serves: 4 to 6
Time: 1¼ hours

1 pound beef (top of round)	3 tablespoons prepared mustard
2 tablespoons oil	1 teaspoon salt
½ cup onions (chopped)	¼ teaspoon cayenne pepper
1 clove garlic (minced)	1 teaspoon chili powder
2 tablespoons green pepper	2 cups canned tomatoes
(chopped)	2 cups canned kidney beans

Have butcher grind the beef. Place oil in a large heavy pan. When quite hot, sauté the onions, garlic, and green peppers until tender. Add the chopped meat and stir well. When slightly cooked, add mustard, salt, cayenne, and chili powder. Mix well and add the tomatoes. Cover and simmer for 45 minutes. Add the beans with their liquid and continue cooking uncovered for 15 minutes.

⟋ *Serve with fluffy rice and cold beer.*

JUICY HAMBURGERS—BROILED

Serves: 6
Time: 4 minutes each side for rare

2 pounds sirloin (ground)	½ cup milk
few douses of seltzer	2 eggs (beaten lightly)
3 white onions (finely grated)	salt and freshly ground pepper
1 clove garlic (crushed)	

Bring steak to room temperature and spread out in large bowl. Preheat broiler. Mix other ingredients well and pour over meat. Let stand for about 5 minutes, then mix lightly with 2 large forks. Coat hands with oil (or cold water) and form meat into 6 large patties. Place in pan under broiler and cook according to doneness you wish.

⟋ *Serve with broiled tomatoes, baked potatoes, and a dry red wine.*

CREAMY HAMBURGERS—SAUTÉED

Serves: 6
Time: Optional

2 pounds chuck steak (ground) ½ cup milk
¼ pound butter 1 cup sour cream
1 large onion (diced) 1 egg
1 clove garlic (minced) salt and freshly ground pepper
8 soda crackers (crumbled)

Place steak in large bowl and allow it to come to room temperature. Heat half of butter in large fry pan and sauté onion and garlic until golden brown, then remove from heat. Meanwhile, soak crackers in milk in small bowl, then drain. Beat egg in sour cream in another. Now add both of these mixtures to meat as well as the cooled onions and garlic (from which butter has been drained). Sprinkle with salt and pepper and mix lightly. This mixture should stand for a while so that all of the ingredients can be blended. Shape into patties—either 6 large ones or 12 smaller ones—and plop them into the same large fry pan with remaining butter (add more butter if needed). Cooking time depends on size of patties and the amount of doneness you desire.

✦ *Serve with stewed tomatoes, baked potatoes, and a dry red wine.*

BURGER-FURTER

This was especially designed for a party of kids.

Serves: 4
Time: About 12 minutes

1 pound sirloin (ground) salt and freshly ground pepper
1 egg (beaten) sprinkle of garlic powder
½ cup cracker crumbs ¼ cup milk
1 small onion (finely chopped) 4 frankfurter rolls (toasted)

Preheat oven (450° F.). Place sirloin into a mixing bowl and gradually add the rest of the ingredients (except rolls). Mix lightly. Now, with hands dampened with cold water or oil, separate the meat into 4 even portions and shape into 4 rolls (resembling frankfurters). Place them in a flat pan and bake for 12 minutes. Immediately put them in the center of toasted rolls.

⸪ *Serve with all of the accompaniments kids like: mustard, ketchup, relish, sliced onions.*

CHOPPED STEAK LAYERED WITH ONION AND PEPPER RINGS

Serves: 4
Time: About 8 minutes

1½ pounds chopped steak
1 egg
2 tablespoons water
salt and pepper

1 large red onion (4 slices)
1 green pepper (4 rings)
cooking oil
soy sauce

Place chopped steak (this could be sirloin or top-of-the-round) in large bowl. Beat egg with water and add. Sprinkle with salt and freshly ground pepper. Combine this mixture lightly with 2 forks and let stand for ½ hour (or more). Preheat broiler. Meanwhile, match up onion slices with pepper rings. They should fit neatly, so discard outer layers of onion if necessary. In a cold, shallow baking pan, place enough oil to barely cover the bottom. Now mold 8 patties lightly from meat mixture. Turn these around in oil so that they have a very slight gloss. Place an onion and pepper ring on top of 4 of the patties. Then place another patty on top. Sprinkle with soy sauce and place under broiler for 3 to 4 minutes. Turn carefully so that onion and pepper rings stay in place. Add a short douse of soy sauce and broil for another 3 to 4 minutes for medium-rare.

⸪ *Serve with baked tomatoes, green salad, and claret wine.*

ARKANSAS MEAT LOAF

Former Arkansas Governor Winthrop Rockefeller loves good food, especially this meat loaf that his wife created. It is highly seasoned and demands a basting sauce. For this reason it stays moist, and should any of it be left over, it is excellent for sandwiches the next day.

Serves: About 6
Time: 1¼ hours

1 pound beef (ground)
1 pound sausage meat (highly seasoned)
1 12-ounce package cottage cheese
1 8-ounce container sour cream
2 slices rye bread (crumbled)
3 eggs (beaten)
2 packages onion-soup mix (Lipton's)
1 tablespoon A-1 sauce

dash of tabasco
Worcestershire sauce (to taste)
soy sauce (to taste)
curry powder (to taste)
dash of powdered ginger
pinch of nutmeg
pinch of rosemary
freshly ground pepper
¼ pound butter
2 onions (chopped finely)
1½ cups chili sauce

Preheat oven (250° F.). Bring meat and cheese to room temperature so that they will be easy to combine. Place these in a large mixing bowl and add the sour cream, crumbled bread, eggs, and onion-soup mix. Stir well, then add the seasonings. (Mrs. Rockefeller is what we call a "natural" cook—she does not measure seasonings, just adds and tastes. This recipe may seem difficult to the novice, but it is a good test for your taste ability.) After you have mixed everything, place in a large loaf pan. Meanwhile, start on the sauce. Melt the butter in a frying pan and add the finely chopped onions. When they are limp and slightly golden, gradually add the chili sauce. Place the meat loaf in the oven and bake for 1¼ hours, basting every 15 minutes with the chili sauce.

ϝ *Mrs. Rockefeller likes to serve this dish with fluffy rice sprinkled with slivers of water chestnuts. Then she usually has a tossed salad (or a green vegetable) and a dry white or a light dry red wine.*

MEAT LOAF PARMIGIANA

Serves: 4
Time: About 30 minutes

8 slices leftover meat loaf
1 cup marinara sauce

½ pound mozzarella cheese
(sliced)

Preheat oven (375° F.). All you have to do is to place the meat loaf slices—in a single row—onto a shallow baking pan. Spoon the sauce over meat and top with cheese slices. Bake for 20 to 30 minutes, or until the cheese has melted.

⸰ *Because this is a leftover (but delicious), it should have some exciting things to go with it. How about an artichoke for each person with a melted butter and lemon dip? Also, chilled rosé wine.*

BEEF AND VEGETABLE SOUP

Serves: 6
Time: About 2 hours

4 rib bones (with ½ inch of
    meat)
salt and freshly ground pepper
6 carrots (chopped)

1 parsnip (chopped)
1 celery knob (chopped)
2 stalks celery (chopped)
1 package dry soup mix

Place bones in a 3-quart pot and fill with 1½ quarts water and add salt and pepper. When water comes to a boil, add carrots, parsnip, celery knob, celery, and soup mix. Cover and simmer for about 2 hours. Remove bones. When they are cool enough to handle, cut off meat and return to soup. Discard bones.

⸰ *This is almost a meal in itself, so the accompaniments should be slight. Try toasted garlic bread and a large salad. Beer or wine would be a fine addition.*

BRAISED FLANKEN (SHORT RIBS OF BEEF)

Serves: 4 to 6
Time: 2½ to 3 hours

4 pounds short ribs of beef (flanken)
2 tablespoons flour
salt and freshly ground pepper
2 tablespoons oil
2 cloves garlic

4 small white onions
3 carrots (chopped in quarters)
1 stalk celery (quartered)
½ green pepper (sliced)
1 cup dry red wine

Preheat oven (300° F.). Wipe ribs off with paper towels, then sprinkle well with a mixture of flour and salt and pepper. Heat oil in Dutch oven and first brown garlic, then push aside and brown flanken on all sides. Remove from heat, add vegetables, and pour in wine. Cover pan and cook for about 2½ hours. Baste frequently and add more wine if sauce has evaporated.

✦ *Serve with noodles, broccoli, and a good dry red wine.*

STANDING RIB ROAST

Serves: 8
Time: 1½ hours (for rare)

4-bone rib roast (cut 4 inches below rib eye)
2 tablespoons oil

1 clove garlic (crushed)
salt and freshly ground pepper

This is also called prime roast of beef. Place roast in a shallow baking pan. Heat oil in small saucepan and add garlic and salt and pepper. Mix well, brush over roast, and let stand at room temperature for 2 hours. Preheat oven (450° F.). Roast for 10 minutes in uncovered pan, then reduce heat to 350° F. and continue for 1 hour and 20 minutes—for rare. Baste with pan drippings from time to time. Roast 15 minutes more if you want a medium roast, another 15 minutes for well-done.

✦ *Serve with browned potatoes and carrots (cooked in pan drippings the last half-hour) and Burgundy wine.*

SHELL ROAST NEW YORKER

Serves: 8
Time: 1¼ hours for rare

8-inch cut of shell roast
1 tablespoon butter (softened)

salt and freshly ground pepper
1 tablespoon chopped onion

Have butcher place 4 1-inch strips of fat on top of the roast—about 2 inches apart—and tie loosely to hold in place. Smooth butter over the meat sections and sprinkle with salt and pepper. Let stand at room temperature for an hour or more. Preheat oven (350° F.). Sprinkle fat with chopped onions and place roast on rack in uncovered roasting pan. Roast for 1¼ hours, if you wish it rare, basting with pan drippings from time to time. Add 15 minutes cooking time for medium, another 15 minutes for well-done.

⌁ *Note: A shell roast is excellent when it is boned. In that case, roast for only 1 hour for rare.*
⌁ *Serve with stuffed baked potatoes, buttered asparagus, and claret wine.*

ROUND ROAST OF BEEF
WITH VEGETABLES

Serves: 8
Time: 2½ to 3 hours

5 pounds eye-round roast
1 tablespoon cooking oil
salt and freshly ground pepper

4 carrots (thinly sliced
  lengthwise)
2 large onions (sliced)
4 large cabbage leaves

Preheat oven (350° F.). Brush oil on roast and sprinkle with salt and pepper. Place carrot strips around sides of roast and tie in place. Cover bottom of roasting pan with onion slices. Place roast in pan and cover it with cabbage leaves. Roast uncovered, 2½ to 3 hours, depending on desired degree of doneness, basting occasionally.

⌁ *Serve with a hearty tossed salad and a dry red wine.*

SAUERBRATEN

Serves: 8
Time: About 3½ hours

4 pounds rump beef
1 bay leaf
4 medium-size onions (sliced)
salt and freshly ground pepper
1 tablespoon brown sugar

2 cups wine vinegar
2 cups water
3 tablespoons flour
2 tablespoons oil
1 cup sour cream

Place beef in large earthenware bowl. Add bay leaf and onions and sprinkle with salt and pepper and sugar. Add vinegar and water (if this is not enough to cover, add equal quantities of each). Cover bowl and refrigerate overnight or longer. Next day, drain off marinade and reserve. Dry meat and onions with paper towels and dredge with flour (save flour). Place meat in Dutch oven sizzling with oil and brown well on all sides. Then add onion slices and brown slightly. Add the marinade to Dutch oven, cover, and cook for about 3 hours. Now remove meat and strain sauce through a sieve. If it seems too watery, mix a little of the flour into drippings left in Dutch oven and gradually add the sauce. Add the sour cream, stirring constantly until the mixture is heated through; do not let it boil. Place meat into sauce, then cover and remove from heat.

⅋ *Serve with Brussels sprouts in butter sauce, noodles, and a husky Burgundy wine.*

RUMP ROAST OF BEEF

Serves: 6 to 8
Time: 3 hours

4 pounds rump of beef
3 marrow bones
2 large onions (chopped)
salt and freshly ground pepper
1 cup carrots (chopped)
½ cup turnips (chopped)

2 cloves garlic (minced)
3 cups Burgundy wine
1 teaspoon thyme
flour
4 tablespoons oil
1 package frozen peas

Place the beef in a large bowl and add all of the ingredients (up to flour). Mix well and refrigerate overnight. The next day, drain and dry the beef, reserving the vegetables and sauce. When beef has come to room temperature, sprinkle it well with flour. Place the oil in a Dutch oven. When it sizzles, brown the meat on all sides. Now scoop out 1 cup of the wine sauce and pour over meat. Cover with a tight lid and cook for 1½ hours over low heat. If sauce has diminished, add a bit more of the wine marinade. Simmer for about 45 minutes, then add the carrots and turnips. About 30 minutes later, add the rest of the marinade and the frozen peas. Leave the pan uncovered and cook for about 15 minutes more. The sauce should be just great at this time. However, if it is too juicy, cool some of the sauce and add a bit of arrowroot (the amount depends on the thinness of the sauce). Stir into sauce in pan.

⁄ *Noodles would be excellent with that fine gravy. So would room-temperature Burgundy wine.*

FAMILY POT ROAST

Serves: 6 to 8
Time: About 3½ hours

5-pound pot roast
3 tablespoons oil
5 large onions (sliced)
2 green peppers (sliced)

4 cloves garlic (crushed)
1 can beef bouillon
salt and freshly ground pepper

Wipe pot roast with paper towels. Heat oil in Dutch oven and add onions. When they are golden, remove onions and brown roast on all sides (about 10 minutes). Return onions to pot and add peppers, garlic, bouillon, salt and pepper. Reduce heat and simmer covered for about 3 hours or until a 2-pronged fork goes into meat easily. Stir once in a while and add a little water if sauce lessens. When meat is done, remove from pot. Let cool and then slice. Meanwhile, place sauce in blender for a few seconds. It should be fairly thick, but if not, mix a little flour with some of the juice from the sliced meat and add to sauce. Return meat and gravy to Dutch oven and let simmer for about 20 minutes.

⸫ *Serve with boiled potatoes, the classic accompaniment. Then something crisp and green—a salad with a simple oil and vinegar dressing—and a dry red wine.*

BEEF STEW WITH WINE

Robert H. O'Brien, chairman of the board of MGM, is also a V.I.P. where cooking is concerned. But if you try to pin him down for an exact recipe, he says, "I just like to fiddle around . . . a good meat, of course . . . then a little of this and a little of that . . . tasting all the way." Then he says, "Ask Leon Lobel, he will give you my accurate recipe." Here it is, just the way Mr. O'Brien cooks it.

Serves: 6 to 8
Time: 2½ hours

5-pound bottom round
¾ cup tomato sauce
8 peppercorns (crushed)
2 cloves garlic (crushed)
½ teaspoon coarse salt
pinch thyme
1 teaspoon dry minced onion

1 teaspoon green pepper
  (finely chopped)
1 tablespoon flour
4 large onions (cut in eighths)
4 carrots (cut in 6 pieces each)
1 green pepper (chopped)
1 teaspoon salt
¾ cup dry red wine

Place the bottom round on a large platter. In a bowl, combine the tomato sauce, peppercorns, garlic, salt, thyme, onion, and green pepper. Mix thoroughly, then add the flour to thicken the mixture. Spread this paste on all sides of the meat and then cover it with cheesecloth, holding it in place with skewers. Meanwhile, place half of the onions, carrots, pepper, and salt into a deep pot, making a bed. On top of this, place the wrapped meat and cover with the remaining half of the vegetables and salt. Cover the pot and cook over a very low heat for 1 hour. Now mix the ingredients lightly and add the wine. Cover pot and cook for another 1½ hours. During this cooking period, remove cover 3 or 4 times, stir ingredients, and add more wine if necessary. Remove the meat to a platter and take off the cheesecloth covering. Slice the meat. Strain the sauce and reheat. Should it seem a bit too liquid, add a little flour mixed with water. Serve this sauce in a gravy boat—it is delicious.

⫩ *Serve with fluffy mashed potatoes and fresh string beans (which have been soaked for 2 hours in white wine before cooking). And, of course, a nice dry red wine.*

TOMATO POT ROAST

Serves: 6 to 8
Time: About 3½ hours

5-pound pot roast	10 peppercorns
3 tablespoons oil	salt
6 large onions (chopped)	8-ounce can tomato sauce
1 bay leaf	8- to 10-ounce can tomato soup

Preheat oven (350° F.). Dab roast with paper towels to take away moisture. Heat oil in large pan or Dutch oven and cook onions until golden brown. Remove onions and reserve. Now brown meat on all sides (about 10 minutes). Discard half of the oil and cover meat with ¾ of cooked onions. Add bay leaf, peppercorns, salt. Cover pan and cook in oven for about 3 hours, or until almost done. Turn roast every ¾ hour and add more water if necessary. Remove meat from pan and slice when cool. Meanwhile add tomato sauce and soup to pan and place over low heat. Place sliced meat on top of sauce and remainder of cooked onions. Cover and simmer for about half an hour.

⫩ *Serve with boiled potatoes and string beans. As for wine, Burgundy would be excellent.*

BRISKET OF BEEF

Serves: 8
Time: About 3½ hours

5-pound brisket	8 peppercorns
3 tablespoons olive oil	8 medium mushrooms (thinly
salt and freshly ground pepper	sliced)
4 large onions (chopped)	4 shallots (chopped)
1 small green pepper (chopped)	½ teaspoon tarragon
3 stalks celery (chopped)	

Pat brisket with paper towels. Heat oil in large pan or Dutch oven. Sprinkle meat with salt and pepper and add to pot. Brown on all sides and remove from pan. Now brown vegetables lightly in oil. Reduce heat and return brisket to pot. Add peppercorns, cover, and simmer for about 3 hours. Preheat oven (350° F.). Place meat in a bake-and-serve pan. Skim off fat from gravy and place it in a small skillet, then pour fat-free gravy over meat. To small skillet (with fat), add mushrooms, shallots, and tarragon. When they have become softened from cooking, place them over meat in baking pan and cook for about half an hour.

✦ With such a husky dish, you could almost forget potatoes, rice, or noodles (they would be good, though). Why not just settle for crisp Italian bread, salad, and red wine?

RAGOUT OF BEEF

Serves: 4
Time: About 2½ hours

3 pounds cross rib	salt and freshly ground pepper
4 strips bacon	1 teaspoon marjoram
2 large onions (sliced)	1 large tomato (quartered)
1 green pepper (sliced)	¾ cup Burgundy wine
2 cloves garlic (crushed)	

Have butcher cut beef into 2-inch cubes. Place bacon in

Dutch oven and allow to cook slowly until crisp and light brown. Remove and place on paper towels to drain. Brown meat on all sides in bacon fat. Then add onions and pepper slices and let brown lightly. Turn down heat and add remaining ingredients, stirring well. Cover and simmer for 2 hours or until meat is fork-tender. During this period, stir occasionally and add more wine if needed.

┐ *Serve with noodles, a green vegetable or salad, and Burgundy wine.*

OLD-FASHIONED BEEF STEW

Serves: 6
Time: About 2¼ hours

2 pounds chuck (2½-inch cubes)	1 large onion (thinly sliced)
2 tablespoons flour	4 carrots (chopped)
salt and freshly ground pepper	¾ cup Burgundy wine
3 tablespoons oil	¼ cup water
2 celery stalks (cut in 2-inch pieces)	1 package frozen peas

Dry meat on paper towels. Place flour and salt and pepper in paper bag. Add meat, a few pieces at a time, and shake until coated. Heat oil in Dutch oven and then brown meat on all sides (about 10 minutes). Remove meat and place on paper towels so that all oil is absorbed. Pour off oil and dab pot dry with paper towels, but leave any segments adhering to pot from browning. Now place a layer of meat on bottom of pan, then a layer of mixed raw vegetables (celery, onions, carrots). Repeat these layers and pour wine and water on top. Cover pot and allow to simmer for about 2 hours, or until meat is tender. Add more wine if juices have evaporated. Add frozen peas and continue cooking for 5 to 10 minutes.

┐ *Serve with boiled potatoes, mixed green salad, and Burgundy wine.*

BEEF STEW WITH BEER

In an article in *Look* magazine called "Secret of a Good Stew," Elizabeth Alston gave "good-stew tips picked up from masters of the pot and from cooking butchers like the brothers Lobel." Here are a few: The better the meat, the better the stew. Have meat cut in regular chunks for even cooking. Cook slowly; fast boiling toughens meat. Use herbs or spices to accent flavor, not to dominate it. She also included this excellent recipe.

Serves: 6 to 8
Time: About 2½ hours

3 pounds top round
¼ cup bacon drippings or oil
3 cups sliced onions
2 tablespoons flour
2¼ cups light beer
2¼ cups dark beer
2 allspice

2-inch bay leaf
⅛ teaspoon thyme
3 peppercorns
6 to 8 slices French bread
   (1 inch thick and toasted)
Dijon-style mustard

Have butcher cut beef into 2-inch cubes. In a heavy pot (preferably one that can go to the table), place the bacon drippings or oil. When sizzling, add meat and brown on all sides. Remove the meat and set aside. Add the onions and stir until they are brown. Sprinkle them with flour and gradually add the beer. Reduce heat and boil slowly, uncovered, for 25 minutes. Tie in cheesecloth the allspice, bay leaf, thyme, and peppercorns. Add this to pot with the browned beef. Cover pot and simmer for 1 hour, then uncover for 30 minutes. Remove cheesecloth bundle and discard. Meanwhile, preheat oven (375° F.), and on toasted bread spread 2 teaspoons of mustard per slice. Place these on top of stew, mustard side up. Place pot uncovered in oven for about 20 minutes or until brown.

Serve with noodles, crisp green salad, and chilled ale.

BEAUJOLAIS BEEF STEW

Jacob S. Potofsky, president of the Amalgamated Clothing Workers and a vice-president of AFL-CIO, frequently devotes evenings to one of his hobbies, cooking. He has given us one of his favorite leisurely recipes.

Serves: 2
Time: 3½ hours

1 pound top-of-the-round	paprika
3 medium onions (thinly sliced)	salt
flour	1 cup Beaujolais wine

Ask butcher to cut beef in bite-size cubes. Place sliced onions in bottom of heavy pan or casserole. Dust the meat with flour and sprinkle with paprika and salt. Put meat on top of onions and pour wine over all. Cover pan and place over very low heat. Cook for 3½ hours, but take a look once in a while. If liquid has diminished too much, add a little more wine.

⟋ *Serve with thin noodles, a green salad, and, of course, a slightly chilled Beaujolais wine.*

FLANKEN STEW

Serves: 6
Time: About 3½ hours

5 pounds shin beef with bone	1 large tomato (quartered)
2 bay leaves	6 cloves garlic
1 large onion	1 bunch celery (cleaned)
10 peppercorns	

Have butcher cut beef into 2-inch sections. Place bay leaves, onion, peppercorns, tomato, and garlic into enough water to eventually cover meat in pot. Bring to boil quickly and add meat. Reduce heat and simmer for about 3½ hours or until the meat is practically falling apart. During the last half-hour of cooking, add the celery, which may be cut in quarters to accommodate to the pot. If the sauce is a bit watery, cool a tablespoon of it, blend with arrowroot, and add.

/ *Serve with noodles because of the good gravy. It is so rich that a green vegetable with tartness would be excellent. How about asparagus with lemon butter, or a crisp salad with oil and lemon?*

CRAZY BEEF STEW

Rod Steiger is well known as an Oscar-winning actor, but many of his friends think he should also get a prize for his cooking expertise. Not just a meat-and-potatoes-man, he uses great imagination in his cookery, such as in the following recipe.

Serves: 8
Time: 3 hours

4 pounds boneless sirloin	2 cloves garlic (minced)
12 whole figs	¼ pound mushrooms (sliced)
½ cup flour	1 cup bouillon
salt and freshly ground pepper	½ cup slivered almonds
¼ pound butter	5 tablespoons seedless raisins
3 medium onions (minced)	2 tablespoons cognac

Have butcher cut sirloin into 2-inch cubes. Preheat oven (350° F.). Dry meat on paper towels. Place sirloin and figs into a bag with flour and salt and pepper; shake to coat well. Melt butter in a large skillet and brown the sirloin and figs on all sides. Add the onions and garlic and stir until brown. Place all of these ingredients into a large casserole. Add the mushrooms, bouillon, and almonds. Cover, place in oven, and cook for 2½ hours, stirring 3 or 4 times during this period. Now add the raisins and cook covered for half an hour. Remove from oven and stir in the cognac.

/ *Serve with fluffy rice sprinkled with drops of pineapple juice and a mixed green salad. Mr. Steiger also likes a dry red wine with the meal, and afterward, cheese, coffee, and a liqueur.*

BEEF BURGUNDY

Don Gussow is president and editor-in-chief of Magazines for Industry, Inc. Both he and Mrs. Gussow love good restaurants, but they both like cooking at home as well. Here is one of their specialties.

Serves: 4
Time: 1 hour plus

2 pounds sirloin steak	1 large can sliced mushrooms
2 tablespoons butter	1 cup Burgundy wine
1 can beef broth	2 cups canned onions
salt and freshly ground pepper	2 tablespoons parsley (chopped)
1 bay leaf	

Have sirloin cut into 1-inch cubes. Bring to room temperature. Melt the butter in a Dutch oven and, when sizzling, brown the meat on all sides. Add the rest of the ingredients and bring to a boil. Reduce heat and simmer for 1 hour, covered. For the last 15 minutes, uncover and stir several times. If juice is too thin, increase heat so that excess moisture will evaporate. Remove bay leaf.

, *Spoon the Beef Burgundy over medium-sized noodles. On the side, have a tart mixed green salad and chilled Ravello Caruso Belvedere rosé. With this meal, a simple dessert of stewed fruit drenched in sweet wine is preferred.*

OSSO BUCO

Serves: 4
Time: About 2½ hours

4 2-inch slices shank of beef bone	4 carrots (coarsely chopped)
½ cup flour	1 large green pepper (chopped)
salt and freshly ground pepper	1 cup dry red wine
¼ teaspoon thyme	½ cup water
3 tablespoons oil	½ cup tomato paste
4 cloves garlic	½ rind lemon (grated)
1 large onion (finely chopped)	½ rind orange (grated)
	2 tablespoons minced parsley

The shank pieces, of course, should be dried with paper towels. Meanwhile, mix the flour, salt and pepper, and thyme in a brown paper bag. Then place each piece of shank bone in the bag and shake it well so that it is well coated. Next, heat the oil in a large skillet or Dutch oven and brown the bones on all sides. Remove these to a platter and add the garlic, onion, carrots, and pepper and sauté lightly in the remaining oil. Reduce heat and return the shank bones. Then add wine, water, tomato paste. Cover and simmer for about 2 hours. The shanks should be quite tender by this time, so add the rind from the lemon and orange and stir. If the juice has diminished, add more wine. Stir and simmer for about 15 minutes. Serve with parsley on top.

⸙ *Serve with any sort of pasta, a green salad, and a dry red wine.*

OXTAIL RAGOUT NO. 1

Serves: 4 to 6
Time: About 4½ to 5 hours

5 pounds oxtails
2 tablespoons flour
¼ cup oil
2 cups onions (chopped)
2 cups carrots (diced)
1 cup celery root (chopped)
½ cup turnips (sliced)
½ cup parsnips (diced)

½ cup celery (chopped)
½ cup mushrooms (chopped)
2 cups water
1½ teaspoons salt
2 beef bouillon cubes
⅛ teaspoon thyme
4 parsley sprigs

Have butcher cut oxtails in pieces. Wash in cold water and dry thoroughly. Roll the pieces in flour. In a wide-bottom heavy pan, add the oil and, when sizzling, add the oxtail pieces and brown on all sides. Remove oxtails and reserve. To the pan, add the onions, carrots, celery root, turnips, parsnips, and celery. Cook over medium heat, stirring frequently. Vegetables should be wilted and browned. This should take about 1 hour. Now return the oxtails, add mushrooms and 2 cups of water, also the salt, bouillon cubes, thyme, and parsley. Cover the pan, reduce heat, and allow to simmer for about 3½ to 4 hours. Add more water if the sauce has diminished. Before serving, remove the parsley sprigs and skim off all grease.

⟨ *Serve with a crisp green salad, French bread, and a dry red wine.*

OXTAIL RAGOUT NO. 2

Serves: 6
Time: 3½ to 4 hours

3 oxtails	2 bay leaves
½ cup flour	½ teaspoon marjoram
½ pound butter	½ teaspoon savory
3 large onions (finely diced)	½ teaspoon basil
6 carrots (coarsely chopped)	salt and freshly ground pepper
3 stalks celery (coarsely chopped)	½ cup sweet red wine

Ask butcher to cut oxtails into joints. Dredge these with part of the flour. Melt butter in Dutch oven and brown oxtail pieces on all sides. Remove meat and add onions, carrots, and celery to heated butter. Sauté until limp and slightly brown, then return meat and add bay leaves, marjoram, savory, basil, and salt and pepper. Add enough water to slightly cover concoction. Cover pan and let simmer for 3 hours. The oxtails should be almost tender by this time, so remove them to a platter. Then remove bay leaves and celery and discard. Put rest of sauce into blender for a few seconds. Save and cool 2 tablespoons of sauce to blend with 1 tablespoon of leftover flour. Add this to "soup" in pot until slightly thickened. Return meat and add wine. Allow to simmer for another 20 minutes.

⟨ *Serve with noodles, the classic accompaniment. Then a tossed green salad and a bit of sweet red wine.*

OXTAIL STEW

Harry Reingold, advertising-agency owner, is actually the cook of the family. But his wife, magazine writer Carmel Berman Reingold, likes to add this or that to the menu. At first she balked at the idea of Oxtail Stew, so he reminded her how inexpensive oxtails were. Then she saw him pour in an $8 bottle of wine. Now they both adore it and feel the stew tastes even better the second day.

Serves: 6
Time: About 3 hours

2 2-pound oxtails (cut in serving pieces)
¾ cup flour
2 teaspoons salt
¾ teaspoon freshly ground pepper
½ teaspoon Hungarian mild paprika
6 tablespoons olive oil
2 cloves garlic
1 cup onion (chopped)

12 small white onions
2 cups tomato juice
1 teaspoon Worcestershire sauce
2 cups Burgundy wine
3 sprigs parsley
2 stalks celery
1 bay leaf
1½ cups mushroom caps
3 potatoes (peeled and quartered)

Wash oxtails and dry them well with paper towels. Roll each piece in a mixture of flour, salt, pepper, and paprika and then shake off the excess flour. Heat the olive oil in a deep, heavy casserole (flameproof pottery is preferred) or in a Dutch oven. Sauté garlic and chopped onion until soft, then brown the small white onions on all sides and remove them. Now add oxtails to pot and brown on all sides. Pour off excess oil and add tomato juice, Worcestershire sauce, and wine (be sure the wine is of good quality). Make a bouquet garni by tying together parsley, celery, and bay leaf. Add this and the browned onions to the pot. Bring liquid to a boil, cover pot tightly, and cook over low heat for 2 hours. Add mushrooms and potatoes. Cover pot again and cook for 1 more hour or until meat is tender. Remove and discard bouquet garni. Skim fat off the top. If this is difficult, chill pot in refrigerator so that fat hardens and is easier to remove. Reheat and serve.

⚹ Serve with cucumber and onion salad and a good Burgundy wine. Because this is a very rich dish, end the meal with a tart fruit sherbet (boysenberry is Mrs. Reingold's favorite).

BAKED CORNED BEEF HASH WITH CABBAGE

Serves: 6
Time: About 45 minutes

3 cups corned beef (diced)
4 heaping tablespoons
   chicken fat
4 large onions (finely chopped)

5 large potatoes (cooked and
   diced)
salt and freshly ground pepper
1 cup cabbage (chopped)

Bring corned beef to room temperature. Heat 3 tablespoons of the chicken fat in a large frying pan. Add onions and brown lightly, then push to one side and add the corned beef pieces, browning them well (from 5 to 10 minutes). Preheat oven (350° F.). Meanwhile, use the rest of chicken fat to grease a loaf-size baking pan. Now place meat mixture and diced potatoes in a bowl, mix well, and season with salt and pepper. Place cabbage in bottom of baking pan and add meat and potato mixture. Be sure that top of loaf is smooth. Bake for 25 minutes. Turn off oven and let stand for 5 to 10 minutes.

⚹ Serve with creamed spinach or asparagus. Beer goes well with this.

5

Veal

AS WE ALL KNOW, veal is the meat of calves. And calves, of course, are the offspring of steers and cows. These youngsters naturally have the same basic configuration of their parents.

↗ TYPES OF CALVES

Since we think so highly of Aberdeen-Angus and Hereford cattle in our beef-buying, we stick to the same breeds and quality when we select veal. There are two types of calves, and the difference is in their age and weight, which affects the appearance and tenderness of the meat. The very young are called vealers, the older variety are just plain calves.

VEALERS

This baby animal is slaughtered when it is from eight to twelve weeks old and weighs 150 to 250 pounds. The meat from this infant calf is sometimes called milk-fed veal because it has been fed entirely on its mother's milk.

CALVES

After the eight to twelve weeks of milk-feeding, calves are allowed to eat grass and grain. When they are almost five months old and weigh from 350 to 400 pounds they are slaughtered.

⸍ THE AGING OF VEAL

Because of the delicate quality and tenderness of veal and its lack of fat content, it cannot withstand any great amount of aging. We usually keep a vealer or a calf in our cooler for no more than a single week.

⸍ QUALITY OF VEAL

When you deal with a butcher who handles only prime beef, you are almost certain that the veal he carries will be of equal prime quality. Before we buy a vealer or a calf, we look through about 300 carcasses. We first look at the conformation of the animal. The back should be broad and barrel-shaped. The shoulders must be short and far apart. The neck should be short and thick. And the leg bones must be small and chunky. Now, the actual conformation of the animal has nothing to do with you—you must trust your butcher about such basic elements. But there are certain things that you should look for.

⸍ TIPS ON BUYING VEAL

First look at the color of the meat. Prime veal should be from almost white to a very light pink. The flesh should have a firm, velvety, and moist look (but not watery). The bones should be small in width and fairly soft to the touch. They should be bright red (as though full of blood). The fat covering the meat should be slight and whitish in color.

The physical makeup and chemical composition of veal is about the same as beef. Veal contains more water than does beef, but it has less connective tissue and little fat. Although there is some marbling in veal, you can hardly see it, due to the whiteness of the meat; also, this graining occurs only in the rib or loin sections.

Even though veal is young and tender, it sometimes takes a different kind and a longer period of cooking than its parent, beef. The reason is that veal does not have the fat and the marbling of beef. Because of this lack of inner lubrication, veal must have "moist" cooking, slower cooking. The intricate marbling in beef makes it excellent for fast broiling. But veal needs gentle and tender treatment.

⁄ WHAT TO AVOID IN BUYING VEAL

- ⁄ Very moist (watery-looking) flesh
- ⁄ Meat that is gray or reddish in coloring
- ⁄ Too much outside fat and inner marbling, which means that the calf has been overfed
- ⁄ Bones that are grayish or white in coloring, which means that the calf is too old
- ⁄ Veal that has yellowish outside fat
- ⁄ Any veal that is obviously sinewy
- ⁄ For scallopini—any cut that is not from the leg

BASIC CUTS OF VEAL

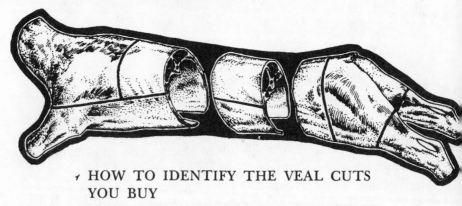

⁄ HOW TO IDENTIFY THE VEAL CUTS YOU BUY

Because vealers and calves are so small, they are not cut into "sides" the way steers are. We buy a complete young animal. Then we hang it in our coolers for no more than a week before cutting it in half lengthwise and later quartering it.

The front and hind sections of calves are named differently from those of steers. For example, the forequarter of beef is called the foresaddle where veal is concerned, and the hindquarter of beef is named the hindsaddle when referring to calves.

Veal is considered the blandest of meats. But as with beef, there are many variations in tenderness and taste.

⸁ HINDSADDLE OF VEAL

This consists of the sirloin, then the rump or leg, and ends up with the hind shank. The tenderest portion is the rump (minus the hind shank). This section makes a wonderful roast that can be complete with bones or boneless and rolled. But it can also be cut into delectable cutlets or thin slices for scaloppine.

CENTER LEG ROAST

This is exactly in front of the hind shank and just before the sirloin. It is an extremely tender roast, with the bone left intact.

ROLLED LEG

This is the same as the center leg roast, but it has been completely boned, rolled, and tied.

SHANK HALF OF LEG

This is an excellent roast. But it is not as tender as the

center leg, as the hind shank is left on, giving it a V-shaped look.

VEAL CUTLET (OR ROUND STEAK)

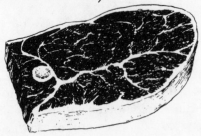

This delicious cut comes from the center of the leg, and the little bone (full of marrow) is left in.

VEAL BIRDS

Actually these are cutlets. They come from the eye of the cutlet and are sometimes called fillet of veal. The bone is removed, then they are pounded and rolled. Sometimes stuffing is involved.

SCALLOPINI

A sketch would not do this wafer-thin piece of meat justice, so we will describe it. It is such a fragile and delicate piece that we only do it to order—never prepare it in advance. We use the intricate French way of advance preparation. First, we dissect the center of the leg, demembrane it, take out any sinewy fibers. When that has been accomplished, we slice to order only. Then we pound it until it is paper-thin or the thickness that the customer desires. This varies from $\frac{1}{8}$ to $\frac{1}{4}$ inch.

⁊ SIRLOIN OF VEAL

This portion is adjacent to the leg. Usually it is left on the leg, but sometimes it is cut off as a separate unit. In this case it is cut into two types of roasts or into superb chops.

STANDING SIRLOIN ROAST

The sirloin of veal is the very next in tenderness to the leg. It is sometimes called a rump roast and the bone is kept in place.

ROLLED DOUBLE SIRLOIN ROAST

This roast is also called a double rump. All bones are removed from the complete sirloin and rump end, then it is rolled into a wonderful concoction that will make easy carving for the host.

⁄ LOIN OF VEAL

The loin section lies right next to the sirloin and is the final cut of the hindsaddle. We sometimes refer to this section as the porterhouse of veal. In a small calf it is only about 8 inches wide and can be cut into small roasts or chops. As the kidney is in this vicinity of the animal, it is frequently part of the cuts.

LOIN ROAST OF VEAL

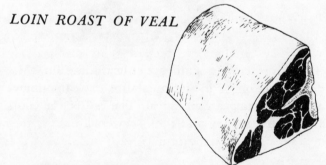

This is V-shaped, like a bunch of chops gathered together. Actually, we often crack the T-bone parts so that after roasting, the carver can cut guest-size portions like a breeze, just like chops.

ROLLED LOIN ROAST

This is actually the same as loin roast, but all of the bones are removed. Before it is rolled, it may be stuffed with veal kidney (even bread stuffing). Then it is rolled and tied with string for baking.

LOIN VEAL CHOPS

The T-bone identifies these chops. They have a large eye and a tenderloin. The tail is usually trimmed off.

KIDNEY VEAL CHOPS

These chops are actually similar to the loins, but they have the added attraction of including a slice of veal kidney. They make a husky serving for any meal. The tail of the chop is wrapped around the kidney and then skewered.

⸱ RIB OF VEAL

Sometimes called "rack," this section comes right after the loin and is the first portion of the foresaddle. The first six bones of the rib are the most tender for either roasts or chops.

RIB ROAST OF VEAL

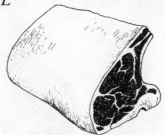

Actually, this roast looks like a series of rib chops. Even though they are not cut from the outside, we usually crack the bone area to make carving as simple as possible.

CROWN ROAST OF VEAL

What a handsome display this roast makes for a small dinner party where the host can carve and serve with the greatest of ease! As you can see from the sketch, the rib chops are "Frenched," then they are slightly cut and cracked at the bone so that they can be bent into a crown shape. Before roasting, the center is filled with an interesting stuffing. Before serving, frills are often added to the top of the bones to dress up the crown and to make for easier eating.

RIB VEAL CHOPS

If you will look back to the rib roast, you will see that these chops are exactly the same shape. So if there is no great demand for roasts, the butcher will cut them into individual chops.

FRENCHED RIB VEAL CHOPS

Although this is the exact same meat as in regular veal chops, it is trimmed quite a bit. The tail is removed from the end bone (as in a crown roast), and these bare bones can be fancied up with frilly paper (as French restaurants like to do it with rib lamb chops).

ꜰ SHOULDER OF VEAL

Even though this forward section of the foresaddle is not as tender as the leg section, it has a lovely taste as a roast (when it has the appropriate amount of cooking). These roasts can be had with the bone left in; or they can be boned with a pocket left for stuffing; or they can be boned, rolled, and tied. This section is more fibrous.

ꜰ NECK OF VEAL

This is a tough section of a calf (and also quite inexpensive). However, it does its duty well for stews, and when boned completely it can be ground for vealburgers.

ꜰ FORESHANK OF VEAL

Although this shank is much more rugged than the hind shank, it has an extremely interesting use. If you have never tasted Osso Buco, you can't imagine what a tough cut of veal (with those good bones left in) can do to your taste buds.

⁊ BREAST OF VEAL

We left this for last because it is not as exciting as the other cuts we have mentioned. It is not exceptionally tender and needs a great deal of cooking. However, because of its bony structure, it can be cut into riblets so that you can have veal spareribs. It also can be made into a long-cooking roast, with bones left in or boneless with a succulent stuffing.

OLD DUTCH WIENER SCHNITZEL

Serves: 4
Time: About ¾ hour

2 pounds veal cutlets	1 tablespoon paprika
½ cup flour	½ cup onions (chopped)
salt and freshly ground pepper	½ cup bouillon
1 egg (beaten)	1 teaspoon lemon juice
1 cup breadcrumbs	½ cup sour cream
3 tablespoons butter	1 tablespoon parsley (chopped)

Cutlets should be ½ inch thick and cut into serving portions. Dredge the meat in flour seasoned with salt and pepper. Dip each piece of meat into beaten egg and then into breadcrumbs. Place meat on platter and refrigerate for an hour or more. Heat butter in skillet over medium heat and stir in the paprika. When bubbling, sauté onions until transparent. Push onions to one side of skillet and add meat, turning it so that it is brown on all sides. Add bouillon, reduce heat, cover pan, and simmer for about half an hour. Now remove veal to a hot dish. Add lemon juice and sour cream to pan. Heat but do not boil. Pour sauce over meat and garnish with parsley.

⁊ *For a truly German touch, serve this with spaetzle (tiny German dumplings). Then you can accompany it with buttered broccoli, a salad, and imported German beer or ale.*

VEAL CUTLET IN WINE

Serves: 4
Time: About 40 minutes

2 pounds veal cutlet
flour with salt and pepper
2 eggs (beaten)
breadcrumbs

2 tablespoons butter
1 cup dry red wine
thyme
1 bay leaf

Preheat oven (350° F.). Cut meat into 4 serving portions and shake in bag filled with flour mixture. Dip each piece in beaten eggs and then in breadcrumbs. Brown quickly in sizzling butter over high heat. Arrange in shallow casserole and add wine, pinch of thyme, and bay leaf. Place uncovered in medium oven for about 30 minutes. If wine dries out, add more to keep bottom of pan moist.

⁊ *Serve with broccoli with lemon butter, noodles, and a dry red wine.*

BREADED VEAL SCALLOPINI NO. 1

Serves: 4
Time: About 8 minutes

4 6-ounce scallopinis (cut
   from leg)
½ cup breadcrumbs
1 tablespoon parsley
2 cloves garlic (crushed)

salt and freshly ground pepper
⅛ cup crushed almonds
1 egg (beaten)
2 tablespoons oil

Spread scallopinis between wax paper and flatten. Meanwhile, mix breadcrumbs, parsley, garlic, salt and pepper, and almonds. Dip each piece of veal into beaten egg, then into crumb mixture. Place between wax paper again, put on platter, and refrigerate for half an hour or more. Heat oil in large skillet over high heat and brown scallopini quickly on each side.

⁊ *Serve with parsley potatoes, stewed tomatoes, and ice-cold beer.*

BREADED VEAL SCALLOPINI NO. 2

Serves: 4
Time: About 8 minutes

4 6-ounce scallopinis (from leg)	3 tablespoons flour
salt and freshly ground pepper	2 eggs (beaten)
1 teaspoon peanuts (finely	2 tablespoons oil
chopped)	8 lemon wedges
1 teaspoon lemon rind (grated)	2 tablespoons parsley (chopped)
3 tablespoons breadcrumbs	

Flatten veal between wax paper. Meanwhile, combine salt, pepper, peanuts, lemon rind, breadcrumbs, and flour. Dip each piece of veal in breadcrumb mixture, then in beaten eggs, and back again into crumb mixture. Place between wax paper, put on platter, and refrigerate for half an hour or more. Heat oil in large skillet. When bubbling, add veal and brown quickly on each side. Drain on paper towels and arrange on hot platter. Garnish with lemon wedges and parsley.

⨍ *Serve with buttered thin noodles, broiled tomatoes, and chilled ale.*

SCALLOPINI OF VEAL MARSALA

Serves: 6
Time: About 10 minutes

2 pounds leg of veal	¼ pound mushrooms (sliced)
½ cup flour	½ cup Marsala wine
salt and freshly ground pepper	¼ cup chicken broth
3 tablespoons butter	paprika

Have butcher cut and pound veal into thin scallopinis. Mix flour with salt and pepper and coat veal on both sides. Heat the butter in a large skillet. When bubbling, add veal and brown quickly on each side. Have hot platter in readiness. As you remove each piece to platter, be sure that you allow butter to drip back into pan. Now add mushrooms to skillet and stir until they become dark and limp. Add wine and broth and boil furiously for about 2 minutes—this will reduce liquid a bit. Pour mushrooms and wine sauce over scallopinis and dust with paprika.

⁊ *Serve with buttered thin noodles, green peas, and a medium-dry red wine.*

INVOLTINI DI VITELLO (STUFFED SCALLOPINI)

The Consul General of Italy for New York, Vieri Traxler, gave us this favorite from the Tuscany region of Italy.

Serves: 6
Time: About 45 minutes

12 slices scallopini	1 egg yolk
salt and freshly ground pepper	1 tablespoon Parmesan cheese
½ pound pork or veal	(grated)
(ground)	strips of lard (fat)
2 slices bread without crust	flour
(dampened)	butter (about ¼ pound)
sprinkle of nutmeg	bouillon

Spread the thin slices of scallopini on wax paper and sprinkle them with salt and pepper. Place the ground meat in a bowl and add the dampened bread, nutmeg, egg yolk, and grated cheese. Mix all of this thoroughly. Spread an equal amount of this paste on each slice of meat. Roll up the slices and skewer them with toothpicks, with a small piece of lard on each side. Dust them with flour. Melt the butter in a large, heavy skillet and brown them on all sides. When nicely browned, add just enough bouillon to cover bottom of pan. Cover and simmer for 25 minutes.

⁊ *Serve with tiny spring peas, pureed carrots, and a dry red Italian wine.*

VEAL SCALLOPINI MOZZARELLA

Serves: 4
Time: About 20 minutes

4 6-ounce scallopinis (from leg)	1 clove garlic (crushed)
2 tablespoons oil	salt and freshly ground pepper
¾ pound chopped veal	½ teaspoon basil
1 egg	½ pound mozzarella cheese
2 tablespoons parsley (chopped)	(sliced)

Preheat oven (450° F.). Place each piece of thin veal gently into a large skillet sizzling with oil. Brown quickly on one side and drain on paper towels. Discard oil and let pan cool. Meanwhile, place chopped veal in mixing bowl and add egg, parsley, garlic, salt and pepper, and basil. Mix well and spread mixture on cooked side of veal pieces. Arrange slices of mozzarella over mixture. Return to skillet and bake in oven for about 15 minutes or until cheese has melted.

⟋ *Serve with buttered string beans and a rosé wine.*

VEAL PARMIGIANA NO. 1

Serves: 4
Time: About 20 minutes

4 large slices scallopini	1 can tomato paste
1 egg (beaten)	1 tablespoon vermouth
seasoned breadcrumbs	Parmesan cheese (grated)
3 tablespoons clarified butter	mozzarella cheese (4 slices)

Preheat oven (350° F.). Dip scallopinis in beaten egg, then in seasoned breadcrumbs. Sauté them quickly on both sides in sizzling butter. Place tomato paste in shallow baking pan and add scallopinis. Rinse tomato can with vermouth and pour over all. Sprinkle Parmesan cheese generously on top. Place in center of oven for about 7 minutes or until tomato sauce is bubbling. Remove from oven and put a thick slice of mozzarella cheese on top of each scallopini. Return to top rack of oven and cook for about 5 more minutes or until cheese has spread over all. If not completely melted or browned, turn oven up to broil and let brown for 2 minutes.

⸗ *Serve with zucchini, an Italian leaf salad (arugula), and a rosé or Italian light red wine.*

VEAL PARMIGIANA NO. 2

Martin Roaman, president of a chain of retail stores, is a gourmet. Although he never goes near the kitchen, he likes to eat what his wife, Carol, devises. They are both sports enthusiasts, and when they have friends over for dinner before hockey or basketball games, she likes to serve dishes that are quick and easy. Here is one of their favorites, and it takes less than half an hour.

Serves: About 6
Time: About 17 minutes

3 pounds veal
3 eggs
1½ cups seasoned breadcrumbs
4 tablespoons salted butter

1 8-ounce can tomato sauce
1 large package mozzarella cheese

Have veal cut and pounded into thin scallopinis. Preheat oven (350° F.). Beat eggs in a large bowl. Meanwhile, place breadcrumbs in large paper bag. Dip each scallopini in egg and then shake in bag with crumbs. Melt butter in large frying pan and brown breaded scallopinis until golden (about 1 minute on each side). Line a large shallow pan with aluminum foil and place veal on this so that pieces do not overlap. Pour tomato sauce over veal and cover each piece with mozzarella cheese. Bake for about 15 minutes or until cheese has melted.

⸗ *Serve with Spanish rice, Caesar salad, and Italian garlic bread.*

VEAL SANDWICHES

Joan Rivers is very funny offstage as well as on (we look forward to her visits to our shop). But she is not at all funny where cooking is concerned. Here is her recipe for what she calls Veal Sandwiches, but her dinner guests feel that this recipe has a cordon bleu aspect.

Serves: 6
Time: About 30 minutes

12 slices veal scallopini
juice of ½ lemon
salt and freshly ground pepper
12 slices Swiss cheese
6 slices ham (paper-thin)
1 cup flour

2 eggs (beaten)
2 cups breadcrumbs
  (unflavored)
3 tablespoons parsley
6 tablespoons butter

Preheat oven (375° F.). Place scallopinis on wax paper and sprinkle with lemon juice and salt and pepper. On 6 scallopinis, place a slice of cheese, then a slice of ham, and then another slice of cheese. Cover this "sandwich" with the remaining scallopinis. There is no need to fasten this sandwich with skewers, as the cheese will melt and hold the veal and ham together. Now dip each portion of meat and cheese into flour, then into the beaten eggs. Meanwhile, combine the parsley and breadcrumbs and dip the sandwiches into this. Because these are quite large, use 2 10-inch skillets and melt 3 tablespoons of butter in each. When bubbling, add the meat and cheese packets and brown well on each side. Then place them in a large shallow pan along with any butter that is left. Put into the oven and cook for about 20 minutes or until the cheese is bubbling. Be sure that the pan is large enough so that there is a separation between the sandwiches to allow for the bubbling of the cheese.

/ This is such a hearty dish that the accompaniments should be quite light . . . perhaps a green salad and a dry red wine.

SALTIMBOCCA (VEAL ROLLS)

Serves: 8
Time: About 40 minutes

8 thin slices milk-fed veal
2 cups Parmesan cheese (grated)
8 slices prosciutto ham
2 tablespoons brandy
4 tablespoons butter
4 tablespoons oil
1 clove garlic (crushed)
½ teaspoon thyme

salt and freshly ground pepper
1 tablespoon tomato paste
¾ cup beef broth
½ cup Marsala wine
2 tablespoons butter
8 large mushroom caps
1 tablespoon lemon juice
2 tablespoons cornstarch
2 tablespoons parsley (chopped)

Place the thin veal slices on wax paper. Sprinkle the cheese over the meat and cover with wax paper. Now pound the cheese well into the meat. Remove top wax paper and place a slice of prosciutto on top of the cheese. Brush on brandy and roll up carefully, securing with string. Place the butter and oil in a large skillet. When sizzling, brown the meat rolls on all sides. Remove the meat and add to juices in skillet the garlic, thyme, salt and pepper, tomato paste, beef broth, and Marsala wine. When well mixed and bubbling, return the meat rolls. Reduce heat, cover skillet, and simmer for about 15 minutes. Meanwhile, melt 2 tablespoons of butter in another skillet and sauté the mushrooms over low heat. Now add the lemon juice to meat sauce and adjust seasoning. Mix the cornstarch with a little water and add gradually to the sauce. Stir constantly and add only enough of the cornstarch mixture to give the correct amount of thickening. Place the meat rolls on a hot platter and remove the strings. Pour the sauce around them and place a mushroom cap on top of each roll. Sprinkle parsley over all.

⌁ *Serve with tiny noodles, a crisp green salad, and chilled beer.*

ROASTED LOIN OF VEAL FINE HERBS

Julia Meade is extremely talented both as an actress and a cook. Here is one of her improvisations.

Serves: 6 to 8
Time: 2½ to 3 hours

1 5- to 6-pound loin of veal (barded)
4 tablespoons parsley (chopped)
2 tablespoons chives (chopped)
1 tablespoon tarragon (chopped)
2 medium mushrooms (chopped)
1 bay leaf (crumbled)
2 medium shallots (minced)
dash grated nutmeg
oil
salt and freshly ground pepper
1 cup beef bouillon (optional)
¾ cup water (optional)
2 tablespoons wine vinegar (optional)

Remove barding from veal and then make a paste of the next 9 ingredients. The amount of oil needed depends on just when the mixture becomes smooth and pasty enough. Also, the amount of salt and pepper will depend on your own taste buds. Rub this mixture all over the loin and then place the fat back on top. Wrap the complete loin in 2 layers of aluminum foil and allow to marinate for at least 3 hours (or overnight) in refrigerator. Before cooking, bring to room temperature—still wrapped in foil. Preheat oven (350° F.). Bake for 2½ to 3 hours or until tender. Place meat on heated platter and remove foil carefully. Discard the fat and scrape the fine herbs into a small saucepan, along with any juice that has seeped out from foil. To this mixture, add bouillon, water, and vinegar, but use your judgment as to how much is needed. The sauce you already have may be juicy enough, so add the extras gradually and mix well and constantly. Then pour the sauce over the meat.

✦ *Serve with grilled tomatoes, green noodles (well seasoned), and a dry white wine. Miss Meade particularly likes the Pouilly Fumé type or the Meursault.*

VEAL ROAST WITH HAM STUFFING

Serves: 8
Time: 2 hours

5-pound shoulder of veal	salt
3 tablespoons white wine	2 slices boiled ham (⅛ inch
3 tablespoons chicken fat	thick)
8 shallots (chopped)	10 thin slices salt pork
1 teaspoon chives (chopped)	

Have butcher make large pocket in veal shoulder. Rub inside and outside of veal with wine and let stand at room temperature for about 1 hour. Preheat oven (450° F.). Meanwhile, melt chicken fat in frying pan and add shallots, chives, and salt. Cook over low heat for about 10 minutes. Remove from heat and cool. Spread this mixture over ham slices, then fold them up and place in veal pocket. Arrange salt pork slices over outside of veal and tie in place with string. Place roast on rack in roasting pan and cook for 15 minutes. Reduce heat to 350° F. and cook for 1½ hours. Baste with pan drippings every half-hour and turn roast on other side for last hour of cooking.

ꞏ *Serve with baked yams, buttered broccoli, and a dry white wine.*

VEAL AND VEGETABLE STEW

Serves: 6
Time: 2 hours

3 pounds shoulder of veal	3 cups water
½ cup flour	1 cup white wine
salt and freshly ground pepper	¼ teaspoon mace
2 tablespoons oil	3 carrots (quartered)
2 cloves garlic (diced)	16 small white onions (peeled)

Cut veal into 1-inch cubes and bring to room temperature, then sprinkle all over with flour seasoned with salt and pepper. Heat oil in large skillet and cook garlic until golden. Push to one side and add veal cubes. When brown on all sides, add water and bring to a boil. Reduce heat and add wine and mace. Cover and simmer for 1 hour. Now add carrots and onions and continue simmering for about 1 hour. Before serving, adjust seasoning. If juice has evaporated somewhat, add a little more wine. But if it seems a bit too watery (which is doubtful), mix a tablespoon of the cooled juice with a little arrowroot and add.

✠ *Serve with boiled potatoes with paprika, chopped spinach, and a medium-dry white wine.*

VEAL AND VEGETABLE CASSEROLE

Serves: 6
Time: 2 hours

3 pounds boneless rump of veal
½ cup flour
salt and freshly ground pepper
4 tablespoons butter
2 large onions (chopped coarsely)
2 cloves garlic (diced)

½ pound mushrooms (sliced)
6 potatoes (peeled and quartered)
1 cup water
1 cup dry white wine
paprika

Have veal cut into 1-inch cubes. Preheat oven (350° F.). Roll cubes in flour seasoned with salt and pepper. Melt 2 table-spoons of the butter in a large skillet over medium heat. When bubbling, add onions and garlic and cook until transparent. Then add mushrooms and stir until they have darkened and are limp. Scrape this mixture into a casserole. Add the rest of butter to skillet and increase heat. Add meat and brown on all sides. Scrape all meat and juices into casserole and add potatoes, water, and wine. Stir well, cover, and bake in oven for about 1¾ hours. After 1½ hours, look at mixture and adjust season-ing. If the sauce seems to have dried out, add a bit more wine. If it is too juicy, leave uncovered in oven for last half-hour of cooking. Before serving, sprinkle with paprika.

✠ *Serve with toasted garlic bread, tomato and escarole salad, and a medium-dry white wine.*

VEAL AND BEEF GOULASH

Serves: 8
Time: About 1¾ hours

2 pounds veal shoulder
2 pounds cross rib or sirloin
1 cup flour
salt and freshly ground pepper
3 tablespoons oil
2 large onions (chopped)
¼ pound Canadian bacon
    (diced)

1 tablespoon paprika
2 large tomatoes (quartered)
1 green pepper (sliced)
½ cup water
1 cup mushrooms (sliced)
1 cup sour cream

Have butcher cut all meat into 1-inch cubes. Roll meat in flour that has been seasoned with salt and pepper. Heat oil in large skillet. Add onions and cook over medium heat until transparent. Push to one side and add meat. Increase heat and brown meat on all sides. Now add Canadian bacon and sprinkle paprika over all, stirring constantly so that paprika blends into juice. Quickly add tomatoes, peppers, and water. When it all comes to a boil, reduce heat, cover, and simmer for 1½ hours. Now add mushrooms and continue simmering for ¼ hour. Adjust seasoning, then add the sour cream. Stir this well into the mixture and allow to heat through, but do not bring to a boil.

⁄ *Serve with wide noodles, buttered turnip greens, and imported ale.*

VEAL AND SOUR CREAM CASSEROLE

Serves: 6
Time: About 1¾ hours

3 pounds shoulder of veal
2 tablespoons butter
¼ cup onions (chopped)
¼ pound mushrooms (sliced)
1 garlic clove (minced)
1 tablespoon flour

½ cup water
1 tablespoon paprika
salt and freshly ground pepper
    to taste
2 cups sour cream

Have veal cut into 1-inch cubes. Preheat oven (300° F.). Heat butter in frying pan over medium heat. When bubbling, add veal and brown on all sides. Remove browned veal and place in a 2-quart baking dish or casserole. Add onions, mushrooms, and garlic to butter remaining in frying pan. If pan has dried out, add a bit more butter. When mushrooms are limp, push to one side and stir flour into butter and gradually add water. When sauce is smooth, add paprika and salt and pepper and cook until thickened—this should take only a minute or so. Remove from heat and stir in the sour cream. Pour this mixture over veal. Cover casserole and bake for 1½ hours.

⸰ Serve with butttered thin noodles with browned sesame seeds on top. To make the serving more colorful, add a green vegetable—asparagus, perhaps. A rosé wine would be a good addition.

DILLED VEAL

Serves: 6 to 8
Time: About 1½ hours

3 pounds boneless shoulder of veal	1 clove
1 pound veal bones	3 tablespoons butter (soft)
4 cups water	6 tablespoons flour
1-inch bay leaf	1 egg yolk
3 sprigs parsley	2 tablespoons lemon juice
3 sprigs dill (or 1 tablespoon dried)	salt and freshly ground pepper
6 peppercorns	3 tablespoons fresh dill (finely chopped)

Have butcher cut veal into 2-inch cubes. Place veal and bones in a 6-quart heavy pot. Add 4 cups of water and bring to a boil, then skim off foam. Meanwhile, tie in a cheesecloth bag the bay leaf, parsley, dill, peppercorns, and clove and add to stew. Reduce heat and simmer, covered, for 1¼ hours. Remove meat to a hot serving dish and cover it. Discard the cheesecloth bag and the bones and allow broth to boil slowly for 15 minutes.

While this is bubbling, use a palette knife to mash softened butter and flour until quite smooth. Take broth away from heat and, using a whisk, beat in this mixture, 1 teaspoon at a time. Replace pot over heat for 3 minutes while you beat together the egg yolk, lemon juice, and salt (1 teaspoon) and pepper. To this mixture, add 3 tablespoons of the hot veal broth. Take away from heat again and gradually add the egg mixture, whisking all the time. Do not return to heat. Just before pouring this sauce over the meat, add the fresh dill (or 1 tablespoon dried).

⸙ Serve with tiny new potatoes, crisp salad, and beer.

SIMPLE VEAL AND GREEN PEPPER CASSEROLE

Serves: 6
Time: 1½ hours

The charm and simplicity of Claire Bloom's acting style are well known to the movie-going public, and the same traits are evident to her friends in her seemingly effortless entertaining and cooking. She likes to be with her guests and therefore prefers long-cooking casseroles that require little of her time in the kitchen.

2 pounds veal	pinch rosemary
salt and freshly ground pepper	2 tomatoes (skinned and
3 tablespoons oil	quartered)
2 green peppers (thinly sliced)	bouillon (optional)

Have veal cut into 2-inch cubes. Sprinkle them well with salt and pepper. In a heavy casserole (preferably enamel), heat the oil and brown veal cubes on all sides. Add the green peppers, rosemary, and tomatoes and stir well. If there is not enough moisture in the pan, add a little bouillon. Cover pan tightly, reduce heat, and simmer for 1½ hours. The only attention you will have to give this casserole is to look once in a while to see if the sauce has diminished and it needs more bouillon.

⁊ *Spoon veal mixture over a bed of rice and have a tossed green salad on the side.*

ALL-AMERICAN VEAL LOAF

A meat loaf is a dish that usually takes a minimum of preparation and little watching. It is a hearty dish for men and popular with children. Here is a classic meat loaf made with veal.

Serves: 6
Time: About 1½ hours

3 pounds neck veal (ground)
1 cup onion soup
½ cup light sweet cream
¼ pound mushrooms (finely chopped)

salt and freshly ground pepper
6 strips bacon (precooked and crumbled)
½ green pepper (finely chopped)

Place ground veal in a large mixing bowl. Meanwhile, simmer onion soup in saucepan for about ¼ hour. Add cream, mushrooms, and salt and pepper and allow to simmer for 5 to 10 minutes, but do not boil. Allow to cool and then add to veal in bowl. Mix well and then refrigerate for half an hour or more. Preheat oven (350° F.). Have a greased 4″ x 8″ baking pan in readiness and smooth the meat mixture into it. It will be slightly rounded on top and look like a loaf of bread. Now, with a knife or your index finger, make crisscross indentations on top of the "loaf." Sprinkle these crisscrosses with the crumbled bacon and green pepper pieces. Bake for 1 hour.

⁊ *Serve with fluffy mashed potatoes and broiled tomatoes.*

QUICK CREAMED VEAL

Serves: 4
Time: About 40 minutes

2 cups cooked veal
1½ tablespoons butter
1¼ tablespoon cornstarch
3 cups veal stock or beef
  bouillon
1 cup heavy cream

⅓ cup almonds (finely
  chopped)
pinch of mace
salt and white pepper
1 tablespoon celery leaves
  (finely chopped)

Have veal cut into half-inch cubes. Heat butter in large skillet. When bubbling, gradually add cornstarch mixed with ½ cup of the bouillon and stir constantly. When this comes to a boil, add veal cubes. Reduce heat and simmer for 5 minutes, then add cream, almonds, mace, and salt and pepper. Stir well and remove from heat for 30 minutes. Add 2½ cups of heated bouillon. Return to low heat and barely simmer, uncovered, for 30 minutes, stirring occasionally. Garnish with celery leaves.

✠ *Serve with fluffy rice, tomato and escarole salad, and a dry white wine.*

6

Lamb

SPRING LAMB is what housewives used to wait for in March and April. Even though they served "regular" lamb on their dining tables from time to time throughout the year, they longed for those juicy morsels of meat that they could get only around Eastertime. And sometimes the first serving of spring lamb involved the superstition of each member of the family making a wish (which was sure to come true), just as with eating oysters on the first day of September. But you can forget those wishes. Because so-called spring (or young) lamb is now yours for the asking in special butcher shops all-year-round, as sheep are raised in every state. And since there are seasonal variations, young lamb is available twelve months of the year. Along with this increase in productivity, there have been great improvements in breeding and feeding methods. As a result, today's lamb has less fat, more protein, and consequently fewer calories.

⁊ TYPES OF SHEEP

When we buy lamb from the wholesale market, the first thing we look for is youth. The younger the lamb the more tender it is, and young lamb has such a delicate flavor and is so easily digested that doctors often recommend it to convalescing patients. When we screen the wholesale market we look for a well-rounded form with light pink meat, white fat, and red-streaked bones. The grade stamp should say prime (or high-grade choice).

HOTHOUSE LAMB

It is only one or two weeks old and has been fed entirely on mother's milk. It weighs a mere 10 to 12 pounds, and when it is cooked you can practically cut the meat with a fork. For large parties it is sometimes cooked whole. But for smaller groups we cut it in halves or quarters.

BABY LAMB

Almost as tender as hothouse, it is four to six weeks old and is most succulent when it weighs 15 pounds, no more than 20. After the first two weeks of mother's milk, it is carefully fed until slaughtering time.

REGULAR LAMB

To be labeled "lamb," the animal must be slaughtered before it is a year old. But most lambs are sent to market by the time they are eight months or younger. We buy nothing over six months, and the maximum weight is never more than 35 pounds.

YEARLING

At one year a lamb becomes a breeder and is called a yearling. The ewe carries her young for five months and then nourishes the baby with her own milk for two weeks. A ewe can produce an offspring once a year.

YEARLING MUTTON

Sheep that are not used as breeders are sometimes slaughtered when they are from one to two years old. Such yearling mutton does not have the delicate flavor or tenderness of lamb. But some people like its somewhat rugged texture and its rather strong taste. After two years, the flavor of mutton becomes stronger and the texture tougher. Only 6 percent of all ovines slaughtered in this country are over a year old. Such mutton is sold to gourmet restaurants that serve mutton chops and to sausage-makers and canning companies. The canners use mutton chiefly for Scotch broth.

⸙ TIPS ON BUYING LAMB

Meat from high-quality young lambs is fine-textured, firm, and lean. It is pink in color and the cross-sections of bones are red, moist, and porous. The external fat should be firm and white and not too thick. In older lambs the meat is a light red, the fat is apt to be thicker and creamy in color, and the bones may look drier, harder, and less red than those of younger lambs.

Lamb does have a degree of marbling. This is hardly perceptible in hothouse or baby lamb but becomes more obvious in older varieties.

⸙ WHAT TO AVOID IN BUYING LAMB

- ⸙ Meat that is dark red
- ⸙ Fat that is yellowish
- ⸙ Bones that are white

BASIC CUTS OF LAMB

⚡ HOW TO IDENTIFY THE LAMB CUTS YOU BUY

As with veal, we buy a complete animal and hang it in our coolers for no more than a week. Since we buy only young animals, their delicacy and tenderness will not withstand any great amount of aging. When a carcass has hung for the time we consider satisfactory, we split it lengthwise and then separate the hindsaddle from the foresaddle. Later we cut it into individual sections—roasts, chops, steaks, and so on.

⚡ THE LEG OF LAMB

This, of course, is the last half of the hindsaddle. The hindmost section is called the shank and the front section the sirloin.

WHOLE LEG OF LAMB

This includes both the shank and the sirloin. It is an excellent roast, but if it seems a bit too large for a customer's immediate needs, the butcher will cut off a few steaks from the sirloin end to be used for another meal.

SHANK HALF OF LEG

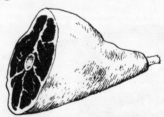

A small and attractive roast that is popular with small families.

SIRLOIN HALF OF LEG

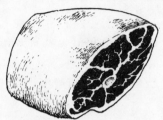

A very tender roast that is excellent for a small group. Or it can be cut into chops, sometimes called sirloin lamb steaks.

ROLLED LEG OF LAMB

In this case the tip end of the shank is cut off. The leg is completely boned and then rolled and tied. Or it can be left open and flattened. This is called "butterfly" leg of lamb.

LEG LAMB CHOPS (OR STEAKS)

These chops are usually quite large and are therefore cut 1¼ inches thick.

LAMB KEBABS

These are delicious little cubes of lamb that are marinated and then strung on a skewer with various vegetables (see p. 139). Although some butchers cut them from other sections of the lamb, we feel that the only "proper" kebabs come from the leg. But they must be dressed first, with all hard membranes taken out.

f LOIN OF LAMB

This section lies at the front part of the hindsaddle or leg. It is very tender indeed and is wonderful for roasts and chops.

ROAST LOIN OF LAMB

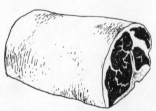

This is tender and delicious and has a small T-bone that separates the tenderloin from the eye. It makes a good party dish, and if there are many guests the complete loin is used. Such a double loin is called a saddle of lamb.

ROLLED DOUBLE LAMB ROAST

Another party favorite because it is so easy to slice. All bones are removed and the saddle is rolled and tied. Sometimes the lamb kidneys are inserted before it is tied up.

LOIN LAMB CHOPS

These are the tenderest chops of all. They have a small T-bone that separates the tenderloin from the eye. They can also have the kidney inserted. In this case it fits neatly below the tenderloin, and the tail is then curved around it and secured with a skewer. When loin chops are boned, trimmed, and rolled (sometimes with bacon), they're noisettes (see p. 137).

ENGLISH LAMB CHOPS

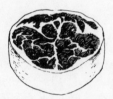

For special dinners, these are considered quite a delicacy. The saddle is simply cut into individual double chops, then the tails are tucked around to form almost a circle of meat that involves two T-bones, two tenderloins, and two eyes.

/ RACK OF LAMB

This section is at the beginning of the foresaddle, right next to the loin. It has no tenderloin but the eye is delicious and tender, both for roasts and chops.

RACK OR RIB ROAST OF LAMB

This is actually a series of rib chops, the number depending on your guest list. The butcher usually cracks the bones so that you can easily carve them into individual chops after roasting.

CROWN ROAST OF LAMB

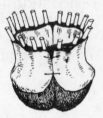

Just as with the crown roast of veal, this makes a handsome display at a party dinner. It is virtually a circle of rib chops that have been Frenched. The butcher has cracked the bones slightly so that the chops can be curved into the classic crown shape, which makes for easy carving and serving. Before roasting, the center is usually filled with stuffing and is cut as a pie.

RIB LAMB CHOPS

These are the same chops that are involved in a rack roast. They have no tenderloin but are delicious. They can be cut into single, double, or triple chops.

FRENCH RIB LAMB CHOPS

What dainty and delectable little chops! The fat is well trimmed, and the end of the bone is left bare. This leaves a little circle of meat attached to a spare bone. For company meals the bone can be dressed up with frilly paper "panties" so that it can be held in the fingers for eating.

⁄ LAMB SHOULDER

Although a butcher once asked a customer if she wanted a back leg or a front leg, there is no such thing in butcher-shop terminology. The front leg of a lamb is always called the shoulder. This section is not as tender as any of those previously mentioned. Some supermarkets use it for roasts with the bone left in or boned and rolled. We use the shoulder for only two types of chops—blade and arm.

BLADE LAMB CHOPS

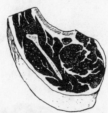

These come from the beginning of the shoulder, right after the rack.

ARM LAMB CHOPS

These chops are cut from the lower part of the shoulder, near the shank, and have a small round bone.

LAMB SHANK AND BREAST

These are cut away from beneath the shoulder and rack in one section and are later divided into smaller cuts.

FORESHANK

As you can see, this portion resembles a tiny roast. However, it is not tender enough for roasting. But it makes an ex-

cellent stew, and when the bone is removed it can be ground
for lamburgers.

RIBLETS

The entire breast cage is a series of ribs. We either cut
them into individual riblets or into sections like spareribs.
These are excellent for barbecuing.

LAMB NECK

The sketch shows just the slices with just a bit of bone,
but there is a long line of bones at the bottom of the neck.
This section is very sweet. Although it is usually used for lamb
stew, it can be boned and then ground. In this case we usually
combine it with the foreshank before grinding. This concoction
makes excellent lamburgers or stuffing for green peppers.

ROASTED SADDLE OF LAMB
WITH PEACH NECTAR

Serves: 6
Time: About 45 minutes

6-pound whole saddle of lamb   12 fresh mint leaves
salt and freshly ground pepper   1 cup peach nectar
1 teaspoon almonds (powdered)

Have butcher score fat in a crisscross pattern. Bring lamb

to room temperature and rub fat with salt and pepper and powdered almonds. Preheat oven (325° F.). Place mint leaves and peach nectar in bottom of roasting pan, under rack. Place lamb on rack and roast for 30 minutes, basting with nectar every 15 minutes, but do not allow leaves to come on top of lamb. Increase heat to 350° F. and continue cooking for 15 to 20 minutes. Baste twice during the last cooking and add more nectar if pan has dried out. Turn off oven and allow lamb to stay in oven for 5 minutes before serving. Baste just before serving.

✓ *Serve with tiny buttered carrots, green peas flavored with mint leaves, and chilled rosé wine.*

ROAST LEG OF LAMB WITH SPECIAL BENNETT PASTE

Joan Bennett is known to her friends as a gourmet and an imaginative cook. Here is one of her favorite dishes for entertaining.

Serves: 6 to 8
Time: 15 minutes per pound

1 leg of spring lamb (6 to 8 pounds)	2 teaspoons Lawry's seasoned salt
1 clove garlic (crushed)	1 teaspoon paprika
2 teaspoons dried ginger	2 tablespoons olive oil
salt and freshly ground pepper	juice from 1 lemon
	1 onion (grated)

Ask butcher to trim off fat from lamb. Mix the other ingredients until they become a smooth paste. Spread this paste all over the lamb and let it stay at room temperature for 2 hours or more. Preheat oven (350° F.). Place lamb in open roasting pan and cook for 15 minutes per pound. During this period a little basting will help.

✓ *Serve with new potatoes in jackets. These should be parboiled for 5 to 10 minutes and then added to the roast 20 minutes before cooking time is over. Miss Bennett likes a liquid mint sauce for the roast, a tossed salad, and a dry red wine.*

GIGOT (LAMB ROASTED FRENCH STYLE)

Serves: 4 (plus)
Time: About 45 minutes

4-pound baby leg of lamb   1 tablespoon garlic (crushed)
1 tablespoon oil

Bring lamb to room temperature and remove a lot of the outside fat. Preheat oven (375° F.). Place oil in roasting pan and, when sizzling, add the garlic. Place fatty side of lamb leg in this mixture until it has become lightly brown. Turn lamb (with fat side up) and place in oven. Roast for about 40 minutes, basting from time to time with pan drippings. By this time the meat should be quite pink, almost red. This is the way the French like it, and it is absolutely tender and delicious.

On French-restaurant menus this dish is often called Gigot aux Flageolets, which means that those delicious little beans should be served with the meat. Cooked and ready to be heated, they are available in many gourmet grocery stores. Aside from that, buttered carrots would be excellent, then an imported dry wine (either white or red).

SPRING LEG OF LAMB—ROASTED WITH CURRANT JELLY

Serves: About 6
Time: 1¼ to 1½ hours

6- to 8-pound leg of lamb   1 large onion (sliced)
salt and freshly ground pepper   6 tablespoons currant jelly
½ teaspoon garlic (crushed)   flour
pinch of oregano   water
10 slices bacon (partially
  cooked)

Ask butcher to trim away all fat from leg of lamb. Rub salt and pepper, garlic, and oregano all over lamb and allow to come to room temperature. Preheat oven (350° F.). Arrange partially cooked bacon slices in a crisscross pattern over leg of lamb and place onion slices on top. Put lamb on rack in a roasting pan and allow to roast for 15 minutes. Remove and add the currant jelly on top of onion and bacon. Continue cooking for 1 hour (longer if it is the larger-size leg). Meanwhile, baste frequently from drippings of pan, adding a bit more water if drippings dry up. At end of cooking, remove roast to a hot platter. Now you can use pan drippings as a "pan gravy"—but you can add a bit of flour (and perhaps a little water) to make a thicker gravy for your gravy boat.

⁊ *Serve with buttered peas, tiny potatoes, and, of course, a scoop of currant jelly. Any type of dry chilled wine will be a charming accompaniment for your guests.*

BUTTERFLY LEG OF LAMB

Serves: 8
Time: About 1 hour

6- to 8-pound leg of lamb
salt and freshly ground pepper
¼ pound butter (melted)
½ cup dry white wine
¼ teaspoon tarragon
1 teaspoon onion (minced)

Have butcher bone leg of lamb, cut open, and flatten to give a butterfly shape. Rub all sides with salt and pepper and bring to room temperature. Preheat oven (475° F.). Meanwhile, combine melted butter, wine, tarragon, and minced onion. Brush lamb with butter mixture and place under broiler and cook for 25 minutes, then reduce heat to 400° F. Turn lamb, brush with butter, and return to broiler for about 35 minutes, brushing with butter mixture throughout cooking. If you like your lamb quite pink, shorten the cooking period. You can judge rareness by making a small gash with the tip of a sharp knife.

⁊ *Serve with tiny potatoes rolled in butter and chopped parsley, chopped spinach, and a chilled dry white wine.*

SPRING LEG OF LAMB—ROASTED WITH HAM

Serves: 6 (plus)
Time: 1¼ to 1½ hours

6- to 8-pound leg of lamb
4 cloves garlic (slivered)
4 tablespoons oil
3 tablespoons lemon juice
oregano
salt and freshly ground pepper

4 slices cooked ham (thinly sliced)
½ cup ginger ale (or slightly more)
3 tablespoons cognac (approximately)

Bring lamb to room temperature and make incisions so that slivers of garlic can fit in neatly. Preheat oven (275° F.). Heat oil in large frying pan and brown lamb on all sides. Now place lamb on rack in baking pan. Dribble lemon juice over it and sprinkle with oregano and salt and pepper. Place the ham slices over lamb and douse with ginger ale. Cover pan and bake for 30 minutes. Increase heat to 350° F. and add cognac. Cook uncovered for about 45 minutes and baste frequently with pan drippings. Should they dry up, add a bit more brandy. The way to tell the doneness of the lamb is to make a tiny incision with the sharp end of a knife.

/ This dish combines the gentle taste of lamb with the saltier one of ham, so the accompaniments should comply. How about tiny buttered potatoes (for the lamb) and broccoli with lemon butter (for the ham)? Of course, a nice dry wine would be fine.

BROILED BABY LAMB WITH SOY SAUCE AND HERBS

Serves: 4
Time: About 1 hour

4-pound baby leg of lamb
oregano

rosemary
soy sauce

Bring lamb to room temperature. Preheat broiler. Make 6 to 8 gashes on fat side of lamb. In these incisions, place crumbled oregano and rosemary. Then douse all over with soy sauce. Place in flat pan amply covered with aluminum foil. Bring ends of foil up over lamb, almost to cover. Place under broiler and cook for about half an hour. Remove pan and turn lamb. Again make incisions and stuff with herbs. Sprinkle soy sauce over lamb, re-cover with foil, and cook again for about half an hour. If you like your meat rosy, cut down the time of cooking on each side.

⁄ *After carving the lamb, baste a bit of the drippings in the aluminum foil over each slice. Then serve with pineapple slices that have been well browned with sugar in the top of the oven, and a green salad in which you have chopped up fresh apples— with a tart dressing, of course. You are on your own as far as wine is concerned. It should be dry and chilled, and can be red, white, or rosé.*

ROAST CURRIED LEG OF LAMB

Joe Namath, the sensational New York Jet quarterback, loves to tackle a good dinner. Two of his favorite dishes are made with lamb and curry. Here is one; the other is on p. 135.

Serves: 5
Time: 1½ hours

1 leg of spring lamb (5 to 6 pounds)	1 tablespoon curry powder
1 clove garlic (slivered)	1 cup water
1 tablespoon rosemary leaves	1 cup rosé wine
½ teaspoon celery salt	6 carrots
½ teaspoon onion powder	6 stalks celery
½ teaspoon coarsely ground pepper	10 new potatoes
	10 small white onions

Preheat oven (500° F.). Place lamb on rack in roasting pan.

Cut pockets in lamb. Insert garlic and rosemary leaves. Sprinkle lamb with celery salt, onion powder, pepper, and rub with curry powder. Add water and wine to pan and place vegetables around lamb. Roast for 15 minutes. Reduce heat to 350° F. and cook for 1¼ hours. Baste every 25 minutes, adding more wine and water if liquid has diminished.

⫏ *This roast, with its built-in vegetables, is a meal in itself, but a crisp mixed green salad would be an excellent addition; so would a slightly chilled rosé wine.*

ROAST SHOULDER OF LAMB WITH APRICOTS

Serves: 6
Time: About 1¾ hours

6-pound shoulder of lamb	1 cup peach nectar
salt and freshly ground pepper	½ cup ginger ale
2 medium onions (quartered)	½ teaspoon ginger
6 dried apricots	

Have butcher crack bones under lamb (to make it easy to carve) and then ask him to tie them in place. Preheat oven (375° F.). Rub lamb with salt and pepper and bring to room temperature on rack in roasting pan. Surround with onions and place apricots on top of meat. Meanwhile, combine peach nectar, ginger ale, and ginger and let stand. Cook lamb for 10 minutes and pour ¼ of peach nectar mixture over top. Reduce heat to 325° F. and continue cooking for about 1½ hours. During this period, pour nectar mixture over roast 3 times (¼ at a time). When roast is done, turn off heat and allow to remain in oven for 5 or 10 minutes. Baste with pan drippings just before serving.

⫏ *Serve with mashed sweet potatoes, buttered Brussels sprouts, and a chilled Sauterne wine.*

DOUBLE RIB RACK OF LAMB
(TWIN LAMB EYES)

Serves: 6
Time: 20 minutes

double rib rack of lamb
6 thin slices lemon
1 tablespoon oil

2 cloves garlic (crushed)
salt and freshly ground pepper

Be sure rack is not cut in half but ask butcher to fillet both sides and French the rib ends. Preheat oven (350° F.). Place 3 lemon slices on each side where the fillet was removed. Mix oil, crushed garlic, and salt and pepper. Brush this mixture on all sides of the fillets and place them on top of the lemon slices. Bake in open pan for 20 minutes. When done, cut fillets in 1-inch slices and return slices to rack and bring to the table on a large hot platter surrounded with vegetables.

/ *Serve with garlic-flavored rice, broiled mushroom caps, buttered carrots, and a chilled rosé wine.*

ROAST RACK OF LAMB
WITH MARMALADE

Serves: 4
Time: About 50 minutes

8-chop rack (or rib) of lamb
salt and freshly ground pepper

½ cup orange marmalade

Bring lamb to room temperature and rub on all sides with salt and pepper. Preheat oven (325° F.). Place lamb bone-side down on rack in shallow pan. Bake for about 30 minutes. Take out of oven and smooth orange marmalade all over, then continue cooking for about 20 minutes.

/ *Serve with broccoli with lemon butter, paprika potatoes, and a chilled Moselle wine.*

ROASTED STUFFED RACK OF LAMB

Richard Avedon is a perfectionist where photography is concerned. When he goes home, he is equally concerned about his palate. This roast lamb is prepared by his excellent cook.

Serves: 6 to 8
Time: About 2 hours

8-pound rack of lamb
½ cup rice (Uncle Ben's)
bouillon
½ pound lamb liver
½ pound lamb kidney
3 tablespoons butter
large yellow onion (minced)
¼ cup slivered almonds
2 tablespoons chopped parsley

1 tablespoon lemon juice
salt and freshly ground pepper
4 tablespoons oil
paprika
1 cup bouillon
1 teaspoon beef essence
 (Bovril)
½ cup sour cream

Bring rack of lamb to room temperature and make the stuffing: First cook the rice in bouillon, in the amount directed on the package for water, for about 20 minutes and set aside. Preheat oven (350° F.). Cut the liver and kidney in small pieces. Place the butter in a large frying pan and add the onion. When transparent, add liver and kidney. Stir this constantly until tender. Remove from heat and add the rice, almonds, and parsley (chopped dill can be substituted for an interesting flavor). Now add the lemon juice and the salt and pepper. Stuff the rack and tie it together so that the stuffing will not fall out. Place 4 tablespoons of oil in the bottom of a roasting pan and place the stuffed rack of lamb on a rack. Cook for about 2 hours, basting every once in a while. During the last half-hour, sprinkle with paprika. Then remove the stuffing to the center of a hot platter. Slice the meat and arrange it around the stuffing. Cover with aluminum foil and keep warm. In the roasting pan, add a cup of bouillon to the drippings, also the beef essence. Bring this to a boil on top of the stove and gradually add the sour cream. Remove from heat, stir well, and pour over meat and stuffing.

⚡ *The Avedons like asparagus with this, and sometimes mashed potatoes with caraway seeds, or tiny new potatoes that have been placed around the roast during the last 20 minutes of cooking. They also like a red wine, Château Bouscaut, Graves 1949, a rare vintage wine.*

STUFFED CROWN OF LAMB

Serves: 8
Time: About 1½ hours

16-chop crown of lamb	3 cloves garlic (crushed)
1 tablespoon lemon juice	10 mushrooms (finely chopped)
2 tablespoons Italian salad dressing	pinch of thyme
	salt
¼ pound butter	1½ pounds ground lamb
1 medium onion (finely chopped)	1 pound chopped sirloin
	1 tablespoon parsley (chopped)

Be sure the butcher has Frenched the bones of the rib chops. Rub outside of crown with lemon juice and inside with Italian salad dressing and bring to room temperature. Meanwhile, melt butter in large frying pan. When sizzling, add chopped onion and crushed garlic. Stir until transparent and then add mushrooms, thyme, and salt. Now add ground lamb and chopped sirloin. Stir well so that the meat absorbs other ingredients. Reduce heat and simmer for 10 minutes, stirring frequently. Remove from heat and allow to cool. Preheat oven (350° F.). Place meat mixture inside the crown of lamb. Pat down firmly and leave a rounded top. Place stuffed crown of lamb on a rack in roasting pan and bake for 1¼ hours. Turn off heat and leave in oven for 10 minutes. Place crown on a large round platter, put paper frills on rib bones, and sprinkle parsley on top of stuffing. Carve like a pie.

⚡ *Serve with buttered boiled potatoes (these may be placed around base of crown), a tossed green salad, and a chilled Chablis wine.*

ROASTED HOTHOUSE LAMB

Serves: 8 to 12
Time: About 1 hour

12- to 16-pound hothouse lamb (whole)
2 tablespoons oil
salt and freshly ground pepper
3 strips bacon (fried and pulverized)
10 stalks celery (coarsely chopped)
8 carrots (quartered)
1½ cups dry white wine
½ cup water
2 tablespoons flour
1 cup bouillon
2 ounces brandy

This type of lamb is only one to two weeks old and must be cooked tenderly (and it must be ordered at least four days in advance). The butcher will cut off ends of legs and shank bones. Make a mixture of the oil, salt and pepper, and pulverized bacon. Rub this all through the inside of the lamb, and if there is anything left over, smooth it on the outside. Bring the lamb to room temperature. Preheat oven (500° F.). Place the lamb on a rack in an open baking pan and roast for 10 minutes. Now place celery and carrots evenly on each side of the lamb, pour the wine over all, and place the water in bottom of pan. Roast for another 10 minutes and reduce heat to 350° F. Continue roasting for 45 minutes to an hour, depending on size of lamb and the degree of doneness desired. This can be tested by making a small incision with the tip of a sharp knife. Be sure to baste frequently with pan drippings during roasting. When you feel the lamb is done, turn off the heat and let pan stand in oven for 10 minutes. Now place the lamb on a hot platter, strain the drippings from the pan, and remove the rack. Put the pan on top of the stove and return the strained drippings. As these are heating, add the flour gradually. When it has become a thick paste (with no flour lumps), slowly add the bouillon. Adjust the seasoning, and at the last moment add the brandy. Serve this in a gravy boat at the table.

╱ *To make this an exceptional dinner-party presentation, cut off the head of the lamb and replace it with a lightly cooked cauliflower. Cover this with Hollandaise Sauce (p. 283) and make a lovely necklace of watercress between the lamb and the cauliflower head. Also, decorate the rest of the roast with watercress. This seems almost enough to serve with such a feast, but should you wish more, try that French accompaniment for lamb, flageolets, a classic go-with for any type of undercooked baby lamb. And, of course, you must have a chilled Chablis.*

CURRIED LAMB CHOPS

Serves: 4
Time: About 50 minutes

8 shoulder lamb chops (1 inch thick)	1 tablespoon curry powder
2 tablespoons butter	1 teaspoon sugar
1 large onion (sliced)	1 tablespoon flour
1 large apple (cored and sliced)	1 cup dry red wine
½ cup seedless raisins	½ cup water
½ teaspoon dry mustard	salt and freshly ground pepper
	celery salt (to taste)
	garlic powder (optional)

Place chops in large skillet in which butter is sizzling. Brown chops on both sides and then remove from pan. Add onion, apple, and raisins and sauté until brown, adding more butter if necessary. Push to one side of skillet and stir in mustard, curry powder, sugar, and flour. When well blended with pan juice, gradually add wine and water until bubbling. Now season to taste with salt, pepper, celery salt, and garlic powder. Return chops to skillet, reduce heat, and allow to simmer until chops are tender.

╱ *Serve with fluffy rice, asparagus with lemon butter, and a light dry red wine (room temperature).*

BROILED LAMB STEAKS

Serves: 4
Time: 10 to 12 minutes

4 lamb steaks (1¼-inch thick)   salt and freshly ground pepper
1 tablespoon oil   pinch of oregano

Place steaks on wax paper. Combine oil, salt and pepper, and oregano and rub on both sides of steaks. Preheat broiler. Let steaks stand at room temperature for half an hour. Place in broiler and cook 5 to 6 minutes on each side.

/ *Serve with broiled tomatoes, baked potatoes, and a dry red wine.*

BROILED DOUBLE LAMB CHOPS

Use 4 double loin or rib chops plus other ingredients for Lamb Steaks and prepare in same way. Only the broiling is different. Place chops in broiler fat side up and cook for 5 minutes or until fat is brown. Turn chops with bone side up and cook for 4 minutes. Now cook each meat side for 2 minutes. In this way the chops are broiled on all 4 sides.

/ *Serve with buttered spinach, creamed onions, and a rosé wine.*

BARBECUED LAMB CHOPS

Serves: 8
Time: About 25 minutes

8 lamb chops (2 inches thick)   1 tablespoon onion (minced)
salt and freshly ground pepper   1 teaspoon barbecue spice
1 8-ounce can tomato sauce   ½ teaspoon dry mustard
2 tablespoons honey   2 tablespoons Southern Comfort

Sprinkle lamb chops with salt and pepper on both sides. Combine remaining ingredients and mix well. Cover chops with sauce and allow to marinate for an hour or more, turning from time to time. Place chops on grill about 7 inches from steady heat. Turn chops several times and brush with barbecue sauce. Cook 20 minutes for medium-rare, 25 to 30 minutes for well-done.

⁄ *Serve with remaining barbecue sauce, potatoes, string beans, and a light red wine.*

NOISETTES OF LAMB

Serves: 4
Time: About 14 minutes

This is a delicious dish. The noisettes are from the tenderest part of a baby loin of lamb, and your butcher will give special treatment to each segment. First he will cut out any bone. Then he will trim off outside fat and curl the meat into a little circle, wrapping the small tail around the eye of the chop. He will wrap a slim slice of bacon around this little bit of roundness and either tie it or skewer it securely.

4 noisettes of lamb (2 inches thick)
4 slices bacon

1 tablespoon bacon fat
salt and freshly ground pepper

Bring noisettes to room temperature. Preheat broiler. Add bacon fat to pan and quickly sauté chops on each side over high heat. Sprinkle with salt and pepper. Now place pan under broiler and cook for about 6 minutes on each side. Be sure that the bacon surrounding the meat is crisp, but the meat should be fairly pink.

⁄ *Serve with buttered peas and pearl onions, currant jelly, and a light red wine.*

BRAISED LAMB SHANKS

Serves: 4
Time: About 1¾ hours

4 lamb shanks
½ cup flour
salt and freshly ground pepper
3 tablespoons oil
1 large onion (chopped)
1 clove garlic (minced)
1 cup bouillon
1 carrot (chopped)

3 stalks celery (chopped)
1 large tomato (peeled, quartered)
½ teaspoon thyme
½ cup Burgundy wine
2 tablespoons chopped parsley
rind from 1 lemon (grated)

Bring lamb shanks to room temperature. Mix flour and salt and pepper and roll shanks in this. Heat oil in Dutch oven and brown meat on all sides. Then add onion and garlic. When transparent, add bouillon and bring to boil. Reduce heat and add carrots, celery, tomatoes, and thyme. Cover and simmer for about 1½ hours. Add Burgundy 15 minutes before end of cooking time, stir well, and leave uncovered. If too juicy, increase heat; if too dry, add more Burgundy. Remove shanks to hot platter and place vegetables and sauce in blender. After blending, reheat sauce, if necessary, and pour over shanks. Garnish with parsley and lemon rind.

⸗ *Serve with fluffy rice, asparagus, and Burgundy wine.*

BRAISED LAMB SHANKS WITH TOMATO SAUCE

Serves: 4
Time: About 2 hours

4 lamb shanks
salt and freshly ground pepper
2 cloves garlic (crushed)
½ teaspoon paprika

¼ teaspoon basil
2 medium onions (sliced)
1 large can stewed tomatoes

Bring lamb shanks to room temperature. Preheat oven (350° F.). Meanwhile, mix salt and pepper, garlic, paprika, and basil. Rub this mixture all around shanks. Place sliced onions in bottom of roasting pan and put shanks on top of them. Cook uncovered for half an hour and then add stewed tomatoes. Continue roasting for 1½ hours. During this period, turn shanks every 30 minutes and spoon tomato sauce over them. Should the tomato sauce seem to dry up a bit, add a little water or some white wine.

⸎ *Serve with parsley potatoes. This will make quite a meal, but you may wish a crisp green salad and a chilled white wine.*

LAMB SHISH KEBAB (OR SHASHLIK *)

Serves: 4
Time: 10 to 12 minutes

2 pounds lamb from leg
  (cut into 20 1½-inch cubes)
1 cup Italian salad dressing
½ cup red wine
8 cherry tomatoes
12 mushroom caps

12 squares green pepper
12 small white onions
1 cup butter (melted)
4 slices cooked bacon
  (crumbled)

Be sure butcher has removed all fat and membranes from meat. Cover lamb cubes with dressing and wine and allow to marinate for an hour or more. Now place tomatoes, mushrooms, green-pepper squares, and onions in a bowl and cover with melted butter combined with crumbled bacon. Let stand for an hour. Preheat broiler. Dry meat on paper towels. Using 4 long skewers, thread meat and vegetables in a colorful pattern. Brush with butter and bacon and broil for 5 to 6 minutes. Turn, brush again with butter and bacon, and broil for another 5 to 6 minutes or until meat has browned.

⸎ *Serve with steaming long-grained rice and a rosé wine.*

* *Shashlik* is the Russian for *shish kebab*.

CURRIED LAMB

Serves: 4
Time: About 1½ hours

1 pound boneless lamb
¼ pound butter
2 large onions (sliced)
2 large apples (peeled, cored, sliced)

1 large can tomatoes
curry powder (amount optional)
salt and freshly ground pepper

Cut meat into 1-inch cubes. Sizzle butter in large skillet and add meat cubes. Brown on all sides. Push to one side and add onions and apples. When onions are golden and apples fall off fork when pierced, add tomatoes (reserving ½ cup of juice). Stir well. Add curry powder to tomato juice and blend well. (Use 1 tablespoon for a mild flavor, more for added spiciness.) Add this mixture to stew and season with salt and pepper. Reduce heat and simmer with cover on for about an hour. Meanwhile, take a look and add water when sauce dries out. Also adjust seasoning at end of cooking.

/ Serve with fluffy rice, chopped peanuts, shredded coconut, chopped orange peel, and chutney. Then have a crisp green salad with a tart dressing. Beer would taste good with this. So would a chilled rosé wine.

LAMB AND KIDNEY BEAN STEW

Serves: 6 to 8
Time: About 2 hours

4 pounds stewing lamb
1 package red kidney beans
2 cloves garlic (crushed)
2 medium onions (diced)

water
salt and freshly ground pepper
1 8-ounce can tomato sauce

Cut lamb into 2-inch cubes and bring to room temperature. Meanwhile, place kidney beans, garlic, and onions in a large pot and add enough cold water to cover. Place over low heat and allow to simmer for about 1 hour, with cover on pot. Now add lamb cubes, salt and pepper to taste, and the tomato sauce. Cover and simmer for about 1 more hour. Take a look now and then and add more water if juice has become too thick.

ı This is such a hearty dish that there is no use in suggesting accompaniments, unless your guests would like some crisp French bread to dunk into that delicious sauce. A green salad with a tart dressing would be a good ending and chilled white or rosé wine a nice go-with.

SAVORY LAMB STEW

Serves: 6
Time: About 2 hours

4 pounds lamb shoulder (boneless)	2 medium onions (chopped)
½ cup flour	1 green pepper (chopped)
salt and freshly ground pepper	1 orange (peeled and sliced)
3 tablespoons oil	1 apple (peeled, cored, quartered)
1 cup bouillon	½ cup apricot brandy
1 cup celery (chopped)	

Have meat cut into 2-inch cubes. Mix flour with salt and pepper and coat meat pieces. Heat oil in Dutch oven, add meat, and brown on all sides. Add bouillon and bring to boil. Reduce heat and add other ingredients (except brandy). Simmer covered for 1¾ hours, stirring from time to time. Half an hour before it is done, add the brandy. If too juicy, increase heat slightly.

ı Serve with wide noodles, mixed green salad, and a rosé wine.

LAMB AND BACON MEATBALLS

Serves: 6
Time: About 30 minutes

2 pounds lamb shoulder
  (ground)
½ pound bacon
1 lemon (thinly sliced)

½ cup milk
½ cup breadcrumbs
1 egg

Place ground lamb in large mixing bowl and bring to room temperature. Preheat broiler. Fry bacon and remove to paper towels to drain. Reserve small part of bacon fat in frying pan and crumble bacon when cool. Place lemon slices in bacon fat and simmer for about 10 minutes. Meanwhile, add crumbled bacon, milk, breadcrumbs, and egg to ground meat and combine lightly with 2 forks. Moisten hands and form meat mixture into balls, either walnut or egg size. Roll them in bacon-lemon sauce and then place in broiler. Cook for about 5 minutes on each side.

✓ *Serve with lyonnaise potatoes, buttered spinach, and a rosé wine.*

LAMBURGERS WITH MUSHROOM CAPS

Serves: 6
Time: About 10 minutes

2 pounds lamb shoulder
  (ground)
12 mushroom caps
4 tablespoons garlic dressing
  (see p. 290)

1 tablespoon soy sauce
1 tablespoon cornmeal (yellow)
2 tablespoons water

Bring ground lamb to room temperature in a large bowl. Preheat broiler. Place mushroom caps in another bowl and add 2 tablespoons of the garlic dressing. Now add the other 2 tablespoons of dressing to the meat along with soy sauce, cornmeal, and water. Mix lightly (but thoroughly) with 2 forks. The seasoning should be just right, but taste a speck of the meat and add salt and pepper if necessary. Form into 12 little lamburgers and place under broiler for 5 minutes. Meanwhile, stir mushroom caps so that they are coated with dressing. Turn burgers and place a mushroom on top of each. Broil again for 5 to 7 minutes.

⸱ Serve with au gratin potatoes, mixed green salad, and a light white wine.

GREEK GRAPE LEAVES STUFFED WITH LAMB

This recipe is from a former cook for Aristotle Onassis and is a favorite of Mr. Onassis's.

Serves: 6
Time: 1½ hours

2 pounds shoulder of lamb
  (ground)
2 cans grape leaves
½ cup uncooked long-grained
  rice
½ cup water
1 medium onion (finely
  chopped)

2 teaspoons fresh dill
  (chopped)
salt and freshly ground pepper
boiling water
½ cup dry white wine
1 teaspoon sugar

Bring lamb to room temperature in a large bowl. Preheat oven (325° F.). Drain juice from grape leaves and spread out individually on paper towels. Meanwhile, cook rice (without salt) until about half-done. Drain rice and, when cool, add to lamb. Mix well and moisten with water. Now add onion, dill, and salt and pepper. When thoroughly combined, stuff grape leaves with meat mixture and roll up. Place in ovenproof dish wide enough to hold stuffed grape leaves in a single layer. Add enough boiling water to barely cover leaves and cook for 1½ hours covered. Half-an-hour before finished, add wine and sprinkle with sugar.

⸱ Serve with sour cream, sautéed eggplant, and chilled Chablis.

LAMB IN PEPPER CUPS

Serves: 4
Time: About 30 minutes

1¼ pounds chopped lamb
2 eggs
2 tablespoons water
1 tablespoon minced onion
salt and freshly ground pepper

¾ cup cooked rice
4 medium-size green peppers
½ cup bouillon
8 pats butter
sesame seeds

Place lamb in large bowl. Preheat oven (325° F.). Add eggs, water, onion, and salt and pepper. Mix lightly with 2 forks and then add rice and mix well. Meanwhile, cut peppers in half and remove all seeds. Stuff meat mixture into pepper cups. Place in shallow pan and surround with bouillon. Cook in oven for 25 minutes, adding more bouillon if bottom of pan has dried out, and baste. Turn oven to broiler heat; remove stuffed peppers, and place a pat of butter on top of each, then sprinkle with sesame seeds. Place under broiler for about 5 minutes or until sesame seeds have browned.

⸱ *Serve with broiled tomatoes, green salad, and a nice dry white wine—perhaps Sauterne.*

SUCCULENT LAMB CASSEROLE
WITH VEGETABLES

Serves: 4
Time: 1 hour and 10 minutes

2 cups cooked lamb
3 tablespoons oil
3 cloves garlic (crushed)
½ cup onions (chopped)
salt and freshly ground pepper
1 8-ounce can tomato sauce

2 tablespoons chopped parsley
2 large zucchini (sliced)
4 carrots (sliced)
2 stalks celery (chopped)
¼ cup Parmesan cheese
   (grated)

Cut leftover lamb into ½-inch cubes. Preheat oven (375° F.). Heat oil in large pan. When sizzling, add garlic and onions and cook until transparent. Reduce heat and add lamb, salt and pepper, tomato sauce, and parsley. Simmer and stir for about 10 minutes. Now place half of the lamb mixture in a 2-quart casserole. On top of this, arrange the zucchini, carrots, and celery. Top this with the rest of the lamb mixture. Sprinkle with cheese and bake for about 1 hour. The vegetables should have made the dish quite juicy. However, if the sauce has dried out, add a bit more water or wine during the last 15 minutes.

≀ *Serve with the potatoes of your choice, a tossed salad, and a medium-dry white wine.*

7

All about Pork—
Yesterday and Today

THE HOG is believed to be the first domesticated animal. Primitive man began to adapt the wild hog to his needs sometime between 7000 and 3000 B.C. Columbus is said to have brought eight pigs with him on his second voyage to the New World, but records indicate that the hog was introduced to America when De Soto landed in Florida with thirteen hogs on board on May 25, 1539. The American Indians liked hog's meat—in fact, they probably originated that old southern favorite, barbecued pig.

Pork was a staple of the Pilgrim diet. The first American meat-packer, a New Englander named William Pynchon, made pork-packing his chief business. Soon Yankee trade with the West Indies sprang up (with "barreled" pork), and before 1700, meat-packers had made Worcester, Massachussetts, a center of export trade.

When the covered wagons went west, hogs went right along with them. On these treks, swine had to forage for food in the forests along the way. Such feeding made for small hams and stringy bacon. But later the settlers fed them with the abundant Indian corn that was readily available, which changed the hog from a scavenger to a standard farm animal. In those days swine grew to be extremely fat animals, which was fine for the housewife, who depended on the hog as her source of fats. But today's homemaker no longer uses or wants as much hog fat—she has become diet-conscious. Therefore researchers and breeders have developed an animal that is much leaner. Today, pork is one of the finest sources of protein and is much lower in calories. For example, an average serving of modern lean pork comes to only 250 calories. Also, pork leads all foods as a source of thiamine (vitamin B_2), and despite any old wives' tales, pork is 96 to 98 percent digestible—one of the highest ratings given any food.

⅋ SWINE PRODUCTION

Hogs are unsurpassed as meat-making machines. A sow can produce two litters each year, and there are usually seven or more pigs in each litter. The United States has the second-largest hog population of any nation, outranked only by China. Nearly four million American farms produce hogs, and about ten billion pounds of pork are processed each year by pork-packers. The old-fashioned pigsty is long gone. Hogs no longer wallow in a muddy enclosure to be fed with leftover scraps and slop. Modern housing for hogs is spick-and-span. The houses are sanitary, heated, and well ventilated to provide protection from drafts, dust, and weather. The bedding is clean and dry. And in the outdoor area, shade is provided. The up-to-date farmer cares for and feeds his swine in a highly scientific manner. He learns from the Department of Agriculture what the nutrient requirements are and feeds them accordingly.

⅋ QUALITY OF PORK AND TIPS ON BUYING

Instead of being graded as prime, choice, and good, the quality of pork is designated by #1, #2, and #3; #1 pork is the best and the only pork sold by butchers who specialize in prime beef, veal, and lamb. The #1 grade of fresh pork has pinkish-gray lean meat with streaks of firm white fat. The meat should be fine in texture and the outer layer of fat should be creamy white and not too thick. The skin should be smooth, free from wrinkles and hair roots. There should be some indication of red (blood) in the bones, signifying a young animal.

ʏ WHAT TO AVOID IN BUYING FRESH PORK

- ʏ Red meat
- ʏ Coarse texture
- ʏ White bones
- ʏ Yellowish fat

ʏ TYPES OF SWINE

SUCKLING PIGS

These are a specialty for holiday parties and must be ordered in advance. They are about three weeks old, weigh about 10 to 18 pounds, and have been fed entirely on mother's milk.

PIGS

Until they are about four months old or until they weigh about 120 pounds, young hogs are called pigs.

HOGS

After four months, swine are on their way to becoming mature hogs. They are usually slaughtered when six months or less. The younger the animal, the tenderer it is. A full-grown live hog weighs about 210 pounds, but this poundage diminishes to 135 pounds when it is divided into retail cuts.

BASIC CUTS OF PORK

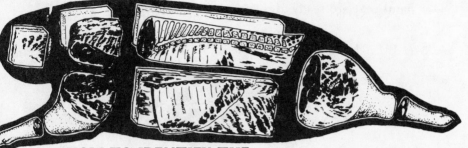

⸗ HOW TO IDENTIFY THE PORK CUTS YOU BUY

In the case of small pigs we buy the whole animal, but where a large hog is concerned, we buy a side or the sections of pork we need—either fresh or smoked.

⸗ LEG OF PORK

The hind section of a hog may be fresh or smoked and can be divided into various cuts or sold whole.

WHOLE HAM

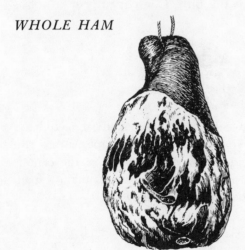

This is a thick, fleshy cut with a high ratio of lean meat to fat and bone. It can be had either fresh or cured and smoked.

Of course, such hams require baking. There are also pre-cooked hams that are ready-to-eat or may be warmed.

HAM BUTT

Because it has little bone, this is the most desirable half of the ham. It can be either fresh or smoked.

HAM SHANK

The end of the ham is rather bony and therefore does not have as much solid meat as the butt. Consequently it is not as desirable as the front half.

HAM STEAK

This is the center slice from a pre-cooked smoked ham, between the butt and the shank. It is oval in shape, with a rim of white fat around one side and a round bone in the center. The meat is lean and fine-grained. Half-inch cuts may be pan-broiled, thicker cuts may be broiled or baked.

⸿ FORELEG (OR SHOULDER)

Like the leg, this section can be bought whole or divided into various cuts. It too may be fresh or smoked.

PICNIC HAM

This is similar to a regular ham both in appearance and flavor. But it has the advantage (for small families) of being lighter in weight. It is available both fresh and smoked.

ARM ROAST OF HAM

This cut is the butt end of the shoulder and is usually sold fresh.

PORK HOCKS

These come from the lower part of the picnic ham, and whether fresh or smoked they make an excellent addition to soups and stews. They contain good-size nuggets of flavorful lean meat.

PIG'S FEET

These, of course, come from either the end of the hindleg or foreleg. They are usually pickled in vinegar. But when bought fresh, they should be simmered until tender (see p. 174) and then served cold in their own jelly.

BOSTON BUTT

This is the upper half of the shoulder. It has a high proportion of lean, tender meat with a uniform covering of clear white fat. It is usually sold as fresh pork.

BLADE PORK STEAKS

When the Boston butt is not sold as a whole roast, it is cut into individual steaks.

BONELESS SMOKED SHOULDER BUTT

Sometimes called "cottage roll," it is a nugget of very lean, rich-flavored meat that is cured and smoked like ham or bacon.

⚡ CENTER SECTION OF HOG

This large portion of the animal is divided lengthwise into two sections that are completely different in character and use. The top center section is the loin which is full of tender, juicy, lean meat. This is usually sold as fresh pork for roasts and chops. Beneath the loin, however, the meat is interspersed with fat (bacon) and bones (spareribs). Much of the lower section is cured and smoked.

SIRLOIN ROAST PORK

Right in front of the butt comes the sirloin. This makes a delicious and tender roast and has a small round bone. It is always sold fresh.

SIRLOIN PORK CHOPS

These are cut from the front part of the sirloin roast.

CENTER LOIN PORK ROAST

Sometimes this is called a rib roast because of the T-bones and rib bones involved. The front part has a tenderloin.

LOIN PORK CHOPS

These delicious chops are cut from the front of the center loin. They have a nice little nugget of tenderloin next to the T-bone.

RIB PORK CHOPS

The meat is similar to the loin chop but there is a longer rib bone and no tenderloin.

ROLLED LOIN ROAST PORK

This is the same as the center loin roast except that the bones have been removed. It has been rolled and tied—so easy for slicing! Here is the way we do it: The bones are taken off in one piece. Then they are placed back onto the roll and tied. The bones are used as a rack while roasting. After the roast has been cooked, the strings are cut and the bones are removed before slicing.

BLADE LOIN ROAST PORK

In front of the center loin, this roast is quite meaty but is marbled with fat. The bone is triangular in shape.

BLADE PORK CHOPS

Cut from the blade roast, this makes a husky and meaty chop. Because of its girth, it should be cooked slightly longer than a loin or rib chop.

SPARERIBS

Even though we have mentioned spareribs from other animals, pork spareribs are a classic. What a delight for barbecuing or broiling! These are the breast and rib bones with tender, juicy, lean meat between them. They come from the lower part of the center section and may be had either fresh or smoked.

They are usually marinated in a fine barbecue sauce (see pp. 170–171), and no one should be concerned about table manners when eating them.

BACON SLABS

Many people have been caught in the trap of buying pre-sliced packaged bacon. You haven't given your taste buds a chance to appreciate flavor until you have bought a complete slab. Such slabs are the side meat from which the spareribs have been removed. If you are not up to carving a slab at home, your butcher will be glad to derind it and slice it to the thickness you desire (a thicker slice than the commercial type can be quite delicious). Of course, all bacon has been previously cured and smoked.

SALT PORK

Small portions of salt pork are used for flavoring such dishes as homemade Boston baked beans and New England clam chowder. Salt pork, as the name implies, has a dry salt or brine cure—it is never smoked.

BAKED VIRGINIA SMITHFIELD HAM

Serves: ¼ pound per person
Time: 20 minutes per pound

Smithfield ham is a great delicacy (expensive too), but if you have never before seen one, you will think it unattractive—somewhat old and moldy. Its looks mean nothing. When it is well cooked, its taste is exceptional. Be sure to slice it very thin, especially when cold. The leftover ham can keep for a number of weeks.

1 Smithfield ham	freshly ground black pepper
cold water (to cover)	brown sugar
2 tablespoons brown sugar	½ cup cider
1 cup cider	½ cup pineapple juice

Soak the ham for 24 to 36 hours in cold water, then scrub it thoroughly with a stiff brush and cold water. Place it in a large heavy pan with skin down and cover with water. Add 2 tablespoons of brown sugar. Bring to a boil and then simmer covered (20 minutes per pound). As water diminishes, be sure to keep the ham covered by adding more hot water. When ¾ cooked add 1 cup cider. Now allow ham to cool in the cooking water. Before ham becomes cold, take it out of water, drain, and remove skin carefully—and avoid tearing the fat. When it is quite cold, sprinkle generously with pepper and cover with a thick layer of brown sugar. Meanwhile, preheat oven to 350° F. and combine ½ cup of cider and ½ cup of pineapple juice, to have in readiness. Place ham in uncovered pan in oven. Baste frequently with the cider–pineapple-juice mixture and let bake until the sugar topping has become a glaze (meaning that the ham has become hot throughout).

⟨ *Serve with candied sweet potatoes and collard or turnip greens— and, because Smithfield ham is such a delicacy, chilled champagne.*

ROASTED SUCKLING PIG

Serves: 10 to 12
Time: About 2½ hours

10- to 15-pound suckling pig	1 large can pineapple chunks
1 cup cognac	1 small red apple
salt and freshly ground pepper	

Have butcher place small piece of wood in mouth to keep open. Place pig on rack in large roasting pan and tuck legs under in sitting position. Then rub inside and out with cognac and allow to stay at room temperature for 2 hours or more, basting with cognac from time to time. Preheat oven (500° F.). Just before roasting, sprinkle pig inside and out with salt and pepper, then surround pig with pineapple chunks and juice. Bake uncovered for half-an-hour. Reduce heat to 350° F. and continue roasting for about 2 hours (basting 5 or 6 times) or until skin is crisp. When pig has cooked for about an hour and ears and tail have become brown, cover them with aluminum foil to avoid burning. Just before serving, remove wood from mouth and insert a red apple.

⸕ *Serve with candied yams, red cabbage, and imported ale.*

BAKED FRESH HAM WITH WHITE WINE

Serves: 12 to 14
Time: 20 minutes per pound

10- to 12-pound fresh ham	2 large onions (coarsely chopped)
¼ pound butter (melted)	3 stalks celery (chopped)
salt and freshly ground pepper	1½ cups white wine
juice of one medium onion	1 tablespoon flour
4 cloves garlic (crushed)	1 cup bouillon

Have butcher score skin and fat of ham. Bring ham to room temperature. Combine butter, salt, and pepper, onion juice, and crushed garlic. Mix well and brush over ham and let

stand for at least an hour. Preheat oven (350° F.). Place onions
and celery in bottom of baking pan. Put ham on top and bake
for 25 minutes. Baste with ½ cup of the wine and reduce heat
to 300° F. and continue roasting until crisp and brown on the
outside and tender when punctured with sharp fork. The entire
cooking time should be about 20 minutes per pound. After first
hour, baste with ½ cup of wine. Baste with pan drippings every
half-hour or so. When done to your satisfaction, remove ham
roast to hot platter. Pour off excess fat from roasting pan and
place pan over medium heat on top of stove. Gradually stir in
flour so that it makes a paste—with no lumps. Add bouillon
slowly and turn up heat. When bubbling, add ½ cup of wine.

✦ Serve with parsley potatoes, slices of cranberry jelly on pineapple
rings, and a medium-dry white wine.

BAKED HAM WITH PINK CHAMPAGNE

Serves: 14
Time: 2 hours

14- to 16-pound precooked smoked ham
1 bottle pink champagne
4 tablespoons orange juice
2 tablespoons pineapple juice
1 tablespoon honey
6 tablespoons dark brown sugar
24 whole cloves
1 large can pineapple rings
maraschino cherries
2 tablespoons peach brandy

Score fat on top of ham in a crisscross pattern, then marinate with champagne overnight in refrigerator. Next day, remove ham from refrigerator and let stand at room temperature
for half an hour. Place on platter and reserve champagne.
Meanwhile, mix orange and pineapple juice with honey and
brown sugar. Rub this liquid all over ham and decorate with
cloves. Place ham on rack in roasting pan and add enough
champagne to cover bottom (about 1 cup). Place in a cold oven
and turn heat to 325° F. Bake for 1 hour. Pour another cup of
champagne over ham, adding more if bottom of pan is not

covered. Bake 15 minutes and baste with pan drippings and champagne. Return to oven and continue cooking for 30 minutes. Now place pineapple rings over ham and put a cherry in center of each. Baste well with pan drippings and additional champagne and cook for 15 minutes. Just before serving, heat brandy and dribble it over all, then flame.

/ *Serve with candied sweet potatoes, buttered cauliflower, and pink champagne.*

BAKED HAM WITH BEER

Serves: 12
Time: About 1½ hours

14-pound smoked ham (precooked)	1 cup brown sugar
1 quart beer	rind of 1 small orange (grated)
3 tablespoons dry mustard	1 teaspoon cinnamon
½ cup port wine	cloves (to decorate)

Ask butcher to score ham in diamond shapes. Preheat oven (350° F.). Place ham in a deep roasting pan and add beer. Cover pan and bake for about 20 minutes (actually, the ham is virtually steaming in the oven heat). Baste the ham 2 or 3 times during this cooking period. Remove ham from roasting pan and allow to cool for about 20 minutes. Meanwhile, combine the mustard, wine, brown sugar, orange rind, and cinnamon. Mix this until it becomes a smooth paste and spread it over the cooled ham, then decorate with cloves in a diamond pattern. Increase heat to 375° F. Place ham on rack on top of beer remaining in roasting pan. Roast uncovered for about 1 hour or until sugar mixture has caramelized (or become brown).

/ *Serve with buttered cabbage, Creamy Mustard Sauce for Ham (p. 283), candied yams, and chilled beer.*

BAKED HAM STEAK WITH PINEAPPLE

Serves: 4
Time: About 25 minutes

1½-pound ham steak   brown sugar
1 can pineapple (4 slices)   cinnamon

Preheat oven (375° F.). Gash fat on edge of steak in several places to prevent it from curling up and place in shallow pan. Arrange pineapple slices on top and pour juice over all. Cook uncovered in oven for 20 minutes. Turn up oven to broiler heat. Remove ham from oven and baste with juices in pan. Sprinkle a little brown sugar over pineapple slices and dust with cinnamon. Place under broiler for 3 to 5 minutes or until pineapple has become lightly brown.

⟋ *Serve with glazed onions, tossed green salad, and a rosé wine*

HAM STEAK WITH PEANUT BUTTER AND GRAPES

Serves: 4
Time: About 25 minutes

1½-pound ham steak   3 tablespoons peanut butter
1 8-ounce can seedless grapes sprinkle of nutmeg

Preheat oven (375° F.). Score fat on outside of ham. Place in a shallow pan and surround with juice from grapes. Cook in oven for 10 minutes. Spread peanut butter on top of ham and baste with juice. Bake for another 10 minutes. Turn up heat to broil. Pour grapes over ham, baste with pan juice, and sprinkle with nutmeg. Place under broiler for about 5 minutes or until juice has been reduced and grapes have a light-brown color.

⟋ *Serve with buttered broccoli spears and a medium-dry white wine.*

BAKED HAM STEAK WITH SHERRY

Serves: 4
Time: About 25 minutes

1½-pound ham steak   12 cloves
1 cup dry sherry

Preheat oven (450° F.). Gash fat on edge of ham diagonally at several points and place in shallow casserole. Pour sherry over ham and sprinkle cloves over top. Cover casserole and cook for 10 minutes. Reduce heat to 325° F. Uncover casserole and cook for about 15 minutes. If it dries out, add a bit more sherry.

✝ *Serve with steamed cabbage wedges with sour cream and nutmeg. (Might be good to have a spicy conserve and beer.)*

HAM STEAK WITH ORANGE

Serves: 4
Time: About 30 minutes

1½-pound ham steak   4 tablespoons pineapple juice
  (center cut)   2 tablespoons orange juice
15 cloves   pinch of cinnamon
1 tablespoon dark brown sugar   4 orange slices
1 teaspoon honey

Score fat on outside of ham and stick cloves into fat. In a cold skillet, mix sugar, honey, pineapple and orange juices, and cinnamon. Place over low heat and stir constantly. When bubbling, place ham steak in mixture, cover, and simmer for 15 minutes. Uncover skillet, turn steak, place the orange slices on top, cover, and cook slowly for 10 minutes. Put steak on a hot platter and let sauce boil for a minute or 2 over high heat. Sauce may be poured over steak or served separately.

✝ *Serve with mashed sweet potatoes, Brussels sprouts, and a rosé wine.*

ROAST LOIN OF PORK

Serves: 6
Time: About 2¼ hours

4-pound loin of pork (bones cracked)
2 large onions (sliced)
1 tablespoon butter
2 apples (peeled, cored, quartered)
½ cup water

Preheat oven (350° F.). Score the fat on top of pork and place fat side down in cold roasting pan. Place pan over medium heat and brown well on all sides. Remove pork to paper towels and brown onion slices in pork fat. Now discard all fat, add butter, return pork, and add apples and water. Cover pan and bake for 2 hours. Uncover for last half-hour of cooking and add a little water if apples have dried out.

⁊ *Serve with buttered cabbage, pickled beets, and ale.*

NORTH CAROLINA ROAST LOIN OF PORK

Mr. Justice Jawn A. Sandifer, of the Supreme Court of the State of New York, is a native of Greensboro, North Carolina, and a connoisseur of southern dishes. He and his wife—an exceptional cook—enjoy this loin of pork roast.

Serves: 6
Time: 3½ hours

2 cloves garlic (slivered)
5-pound pork loin
1 teaspoon thyme
coarse salt
12 small new potatoes
6 carrots (quartered lengthwise)
12 small white onions
white pepper

About an hour before roasting, insert tiny slivers of garlic into the fat of the pork. Then rub the fat with thyme and coarse salt. Preheat oven (325° F.). Place the pork on a rack in a shallow pan and roast for about 1½ hours. Cut a band from the skin of the new potatoes and, along with the carrots and onions, place them around the roast. Sprinkle all with white pepper and return to the oven for another hour. Remove the roast to a hot platter and let it rest for 15 minutes. Arrange the vegetables around the pork before taking it to the table.

⁊ *Serve with apples sautéed in butter and sprinkled lightly with sugar, a tossed green salad, and a chilled dry white wine.*

ROAST LOIN OF PORK—SWEET AND PUNGENT

Serves: 6
Time: About 2½ hours

4-pound loin of pork
  (boneless)
10 dried apricots
15 cooked chestnuts (shelled
  and minced)

3 tablespoons apricot jam
1 tablespoon apricot brandy
2 tablespoons water

Have butcher cut slits about 2 inches apart and 1½ inches deep in fat side of meat. Preheat oven (350° F.). Stuff slits with apricots. Place meat on rack in roasting pan and sprinkle minced nuts over top. Roast meat for 2½ hours. Meanwhile, combine jam, brandy, and water and use it to baste pork after first hour of cooking. Baste every half-hour from then until well cooked. Add more water or brandy if pan drippings seem to dry up.

⁊ *Serve with buttered broccoli, spicy apple sauce, and beer.*

ROASTED CROWN LOIN OF PORK

Serves: 10 to 12
Time: About 3 hours

whole loin of pork
1 tablespoon flour
salt and freshly ground pepper

4 strips bacon
Fruit Stuffing for Crown Loin of
  Pork (pp. 279–280)

Have your butcher trim the loin of pork and form it into a crown, tying it with string. Then ask him to cut away the meat from the top bones so that the roast really resembles a crown. When you get this very attractive cut of meat home, bring it to room temperature. Preheat oven (350° F.). Meanwhile, combine the flour with salt and pepper and rub it over the outer fat of the crown. Then cook the bacon and crumble it. Brush the outside bone with a mixture of crumbled bacon and fat. Fill the center of the crown with Fruit Stuffing and place a bit of aluminum foil on each of the bone tips so that they will not burn. Roast for half an hour in an open pan. Reduce heat to 325° F. and continue roasting for about 2½ hours. You should allow about 25 minutes per pound for roasting. When roast is done, remove the foil from bone ends and add frills that your butcher has supplied.

⸜ *The stuffing is quite filling, so all you will need is a cabbagey-type vegetable—broccoli, Brussels sprouts, turnip greens, or just plain buttered cabbage. And a cool imported ale would be tasty.*

PORK CHOPS WITH PINEAPPLE

Serves: 4
Time: About 50 minutes

4 large loin chops	1 can pineapple (4 slices)
soy sauce	cinnamon

Preheat broiler. Trim fat from chops. Place in large shallow pan. Douse both sides with soy sauce and let stand for 10 to 20 minutes. Put under broiler—5 minutes for each side or until brown. Reduce heat to 350° F. and remove chops from broiler. Place pineapple slices around chops and pour juice over all. Cook for 20 minutes. Turn chops and pineapple and cook for another 20 minutes. Before serving, dust lightly with cinnamon.

⸜ *Serve with glazed carrots, green salad, and a rosé wine.*

PORK CHOPS IN ORANGE MARINADE

Serves: 6
Time: About 45 minutes

6 large loin chops
2 cups orange juice

rind of 1 lemon (grated)
2 tablespoons honey

Trim chops of most fat. In large bowl mix orange juice with lemon rind and honey. Place chops in this mixture, turning them so that they become well saturated. Refrigerate overnight or until ready to use. Bring to room temperature for about an hour, turning several times. Preheat oven (375° F.). Spoon ⅓ of marinade over chops, reserving rest. Place chops in oven for about 45 minutes. Halfway through cooking, turn chops. If they have become too dry, add a bit more of the marinade.

⸙ *Serve with baked beans, buttered turnip greens, and beer.*

PORK CHOPS IN RED WINE

Serves: 4
Time: About 40 minutes

4 large loin chops
1 tablespoon butter
salt and freshly ground pepper
paprika

6 shallots (chopped)
basil
thyme
1 cup dry red wine

Preheat oven (350° F.). Trim fat from chops and place in heavy skillet sizzling with butter. Sauté over high heat until brown on both sides. Sprinkle with salt and pepper and paprika. Place chops in casserole, surround with shallots, and sprinkle with a pinch of thyme and basil. Add wine. Cook covered for about 35 minutes, adding more wine if necessary.

⸙ *Tip: This dish may be "held" longer by reducing heat to low, but be sure there is enough wine to keep moist.*
⸙ *Serve with buttered spinach, spiced peaches, and a dry red wine.*

PORK CHOPS WITH APPLE BRANDY

Serves: 6
Time: About 45 minutes

6 large loin pork chops
2 large apples (peeled, cored, sliced)
1 ounce water
2 ounces apple brandy

⅛ teaspoon mace
¼ teaspoon nutmeg
salt and freshly ground pepper
½ teaspoon honey

Trim fat from chops and let stand at room temperature. In a large cold skillet, place apples, water, and 1 ounce of the brandy. Sprinkle this with mace, nutmeg, and salt and pepper, then stir in the honey. Cover pan and place over low heat. Simmer for 15 minutes. Uncover, push apple mixture to outer edges of pan, and add pork chops. Cook chops over medium heat for 15 minutes, then turn and cook for another 15 minutes. If not brown, increase heat and cook a few minutes longer. Remove browned chops to hot platter. Add second ounce of brandy to sauce, stir well, and pour over chops.

✦ *Serve with buttered broccoli, mashed sweet potatoes, and a chilled Sauterne wine.*

PORK AND SAUERKRAUT STEW

Serves: 6 to 8
Time: About 2 hours

3 pounds boneless loin pork
3 pounds spareribs
4 large onions (chopped)
3 cloves garlic (minced)
1 teaspoon caraway seeds

1 teaspoon celery seeds
salt to taste
1 can chicken broth
2 pounds sauerkraut (drained)
1 pint sour cream

Have pork cut into 1½-inch cubes and ribs into 2-inch sections. Now place pork and ribs in a very large skillet over low heat. Increase heat and brown on all sides. Meanwhile, combine onions, garlic, caraway seeds, celery seeds, salt, and broth and add to meat. Bring to a boil over high heat and then reduce heat and simmer covered for 1 hour. Now add the sauerkraut and mix well. Continue simmering for ¾ hour. Remove from heat and allow to cool for a few minutes, then add the sour cream. When well mixed, return to low heat to warm the ingredients. Do not boil.

♪ *Serve with boiled potatoes sprinkled with paprika, and beer.*

PORK CASSEROLE WITH COGNAC

Serves: 4
Time: About 2 hours

2½ pounds loin of pork (boneless)
¼ cup flour
1 tablespoon crisp bacon (crumbled)
salt and freshly ground pepper
4 tablespoons butter
1 medium-size onion (chopped)
1 clove garlic (minced)

½ pound mushrooms (sliced)
1 small can water chestnuts (drained)
1 cup bouillon
½ cup heavy cream
½ cup cognac
sprinkle of paprika
2 teaspoons crumbled bacon

Have butcher cut pork into 1-inch cubes. Preheat oven (350° F.). Place flour, crumbled bacon, and salt and pepper in paper or plastic bag. Shake pork cubes in this until they are coated on all sides. Melt butter in large skillet. When bubbling, add pork and brown quickly on all sides. Push to one side and add onions and garlic. Cook until transparent, then place with meat in a 3-quart casserole. Reduce heat under skillet and add mushrooms, adding more butter if original butter has been absorbed by meat. Cook until darkened and limp and add to casserole along with water chestnuts and bouillon. Stir well, cover, and bake in oven for 1 hour and 45 minutes. Just before serving, stir in the cream, then the cognac. Garnish with paprika and crumbled bacon.

♪ *Serve with riced potatoes, romaine salad with a tart dressing, and a chilled Moselle wine.*

MEAT AND CHICKEN BARBECUE
(OVEN-COOKED)

Joel Grey is not only a well-known actor; he and his wife are also exceptional hosts. Here is one of their party favorites.

Serves: 8 or more
Time: About 2 hours

5 pounds roast beef bones
3 pounds spareribs (cut in sections)

16 chicken wings
8 chicken legs (with thighs)

Sauce "Florence"

14 ounces ketchup
12 ounces chili sauce
½ cup prepared mustard
1½ cups brown sugar
freshly ground pepper
1½ cups wine vinegar
1 cup fresh lemon juice

½ cup thick steak sauce (A-1)
dash tabasco
¼ cup Worcestershire sauce
1 tablespoon soy sauce
2 tablespoons salad oil
1 12-ounce can beer
liquid smoke (to taste)

Preheat oven (350° F.). Arrange roast beef bones and spareribs in a large baking pan. When oven is hot, place pan in center rack and allow fat to dissolve (this should take about 20 minutes). Meanwhile, wash chicken and dry. Then combine sauce ingredients in a large pot—first put in the ketchup and chili sauce, then stir in the mustard, sugar, and pepper. When this is well mixed, gradually add the vinegar and lemon juice. Place pot over low heat and stir in the steak sauce, tabasco, Worcestershire sauce, soy sauce, and oil. When this mixture is gently bubbling, gradually add the beer. The liquid smoke should be added a few drops at a time so that you can taste and decide how much more to add. This is a very hot and delicious sauce and should be allowed to simmer uncovered for at least 45 minutes. Meanwhile take the baking pan from oven and drain off all the fat that has accumulated. Add the chicken wings and legs and arrange so that there is no overlapping. Brush meat and chicken with sauce until they are well coated. Return pan to oven for 1 hour. Keep barbecue sauce warm and brush meat and chicken every 10 minutes. Place leftover sauce in gravy boat.

ꞁ *Serve with unbuttered rice with barbecue sauce on top,* arugula or Boston lettuce salad, and a dry red wine.

SPARERIBS WITH WINE VINEGAR

Serves: 4
Time: About 1½ hours

4 pounds spareribs
1 pint wine vinegar
1½ cups sugar

3 ounces ginger root (thinly sliced)
2 tablespoons cornstarch
water

Have ribs cut into 2-inch pieces and place in a large cold skillet. Place over low heat at first, then increase heat so that ribs can brown well on all sides. This should take about 10 minutes. Discard all drippings. Meanwhile, combine vinegar, sugar, and ginger root and add to ribs. Bring to a boil, lower heat, and simmer covered for about 1¼ hours. Dissolve the cornstarch in a little water and add gradually to sauce to thicken.

ꞁ *Serve with mashed sweet potatoes, buttered string beans, and ale.*

TEXAN SPARERIBS

Serves: 4
Time: 1½ hours

4 pounds spareribs
2 large onions (chopped)
½ cup ketchup
1 tablespoon Worcestershire sauce
1 tablespoon sugar

¼ teaspoon chili powder
2 stalks celery (chopped)
1 tablespoon wine vinegar
1 clove garlic (crushed)
boiling water

Have butcher cut spareribs into 2-inch rib sections. Put ribs in a large skillet and place over low heat. Gradually increase heat so that they are browned on all sides. Discard all fat drippings. Meanwhile, combine onions, ketchup, Worcestershire sauce, sugar, chili, celery, vinegar, and garlic. Then add enough boiling water to make the mixture the consistency of a light gravy. Pour this sauce over ribs and and let stand for several hours, even overnight. Preheat oven (325° F.). Before cooking, bring to room temperature, then bake in oven for about 1 hour.

↲ *Serve with sauerkraut, small boiled potatoes, and either beer or ale.*

HAM AND SPLIT PEA SOUP

Serves: 6
Time: 2¾ hours

3 pounds smoked ham shank
1½ cups green split peas
9 cups cold water
3 tablespoons butter
¾ cup celery (chopped)
½ cup carrots (chopped)
3 medium onions (chopped)
2 cloves garlic (crushed)
salt to taste
15 peppercorns (in cheesecloth bag)
rind of 1 orange (grated)
2 knockwurst sausages (sliced)

Trim fat from ham shank and place in a large kettle with split peas and water and bring to a boil. Reduce heat and simmer for 1 hour covered. Meanwhile, melt butter in skillet and add celery, carrots, onions, and garlic and sauté for about 15 minutes over medium heat. Pour vegetables into ham pot and add salt, peppercorns, and orange rind. Cover and cook for 1 hour. Then remove ham shank from stew and cut off meat in bite-size cubes. Discard bone and bag of peppercorns. Mash the vegetables and return ham cubes to soup. Add sliced knockwurst and simmer uncovered for 30 minutes.

↲ *Serve with toasted garlic bread, Caesar salad, and ale.*

HAM BONE AND SPLIT PEA SOUP

Serves: 6
Time: About 3¾ hours

¼-pound salt pork (sliced)
1½-pound ham bone
14 cups water
3½ cups green split peas
2 tablespoons barley
2 large fresh turkey wings
¾ cup onions (chopped)

1¼ cups celery (finely chopped)
¾ cup carrots (chopped)
1 15-ounce can tomatoes
½ teaspoon dried thyme
1 tablespoon soy sauce
salt and freshly ground pepper to taste

Place salt pork in large kettle and allow to brown slightly. Add ham bone, water, split peas, barley, and turkey wings. Bring to a boil, then reduce heat, cover, and simmer for 2½ hours. Now add vegetables, thyme, and soy sauce and continue simmering for 1¼ hours. Discard ham bone. Remove and bone turkey wings, chop meat, and return to soup. Discard bones. Cool, then refrigerate overnight. Next day, skim off top fat, stir well, and add salt and pepper if needed. Reheat and serve hot.

/ This is such a hearty soup that it is truly a meal in itself. All that's really needed as an accompaniment is a generous salad bowl —a combination of tomatoes, greens, and scallions with a sharp French or Italian dressing. Add crunchy hot rolls and sweet butter and you have a delicious meal.

HAM HOCKS AND COLLARD GREENS

Serves: 8
Time: About 3 hours

4 ham hocks
6 pounds collard greens (chopped)

1 red pepper (chopped)
2 onions (halved)
salt (about 2 teaspoons)

Cover ham hocks with cold water and bring to a boil. Reduce heat and simmer for 2 hours or until tender. Remove hocks and place in a hot covered dish. Place collard greens in the ham water and add the red pepper, onions, and salt. Cover pot. Simmer for about 1 hour. Uncover for last half-hour so that liquid diminishes. Just before serving, place ham hocks on top of greens and let them heat through.

⸗ Serve with baked yams, hush puppies, and a chilled ale.

KENTUCKY HUSH PUPPIES WITH BACON

"Soul food" is so popular these days that we did not wish to leave out one of the most southern of all dishes—hush puppies. Our problem was that this is strictly a meat cookbook, so we found a friend who added the bacon. Even though she usually makes them without bacon, she has found that they are very tasty with. She serves these as hors d'oeuvres or as an accompaniment to other "soul" foods.

Serves: About 8
Time: About 15 minutes

6 slices bacon (cooked and crumbled)
2 cups cornmeal
1 tablespoon flour
½ teaspoon baking soda
1 teaspoon baking powder

1 teaspoon salt
3 tablespoons onion (grated)
1 cup buttermilk
1 egg (beaten)
oil

Cook the bacon until crisp. Remove to paper towels to drain, crumble, then discard bacon fat. Now sift together the dry ingredients and add the onion, buttermilk, and beaten egg and mix well. Meanwhile, heat oil (1½ inches deep) in deep fry pan. You will know it is hot enough when a small piece of bread browns quickly in it. Drop the dough into the oil with the spoon size you wish and allow to become golden brown. As the Hush Puppies are done, remove them and drain on paper towels. If you are serving them as a side dish for other "soul" food, place a bit of butter on top of each.

ALABAMA PIG'S FEET

Although pig's feet may seem plebeian to some people, they are considered quite a delicacy to many—and not only southerners!

Serves: 6
Time: About 2 hours

6 small pig's feet
3 cloves garlic (crushed)
1 onion (sliced)
8 peppercorns
2 tablespoons salt

Scrub the pig's feet with a strong brush and rinse with cold water. Place them in a heavy iron pot and barely cover with cold water. Add the garlic, onion, peppercorns, and salt and bring to a boil. Reduce heat, cover pan, and simmer for about 2 hours.

↑ Serve with potato salad and cold beer. Be sure to have vinegar handy so that each guest can sprinkle his pig's foot with it.

BROILED EGGPLANT, HAM, TOMATO, AND CHEESE COMBO

Serves: 4
Time: About 20 minutes

1 large eggplant
flour
salt and freshly ground pepper
3 tablespoons cooking oil
4 generous slices ham
2 large tomatoes (each cut into 2 thick slices—ends off)
4 large pats butter
salt, pepper, sugar
mozzarella cheese (4 thick slices)

Cut eggplant into 4 thick slices. Peel off skin and place slices between paper towels with heavy weight on top. Let stand for about an hour (change paper towels if they become too moist). Preheat broiler. Dust eggplant slices lightly with flour seasoned with salt and pepper. Brown them briefly in hot oil. Place slices into shallow pan. Put one ham slice on top of each eggplant slice, then a thick tomato slice. Add large pat of butter on each and sprinkle with salt and pepper and sugar, Place under broiler for 5 minutes. Remove and put a large slice of mozzarella cheese on top of each tomato. Place under broiler again for 5 minutes—or until cheese has melted and streamed down over all.

⸗ *Serve with buttered Italian string beans and a fine rosé wine.*

EGGS BENEDICT

Serves: 2
Time: About 20 minutes

4 slices ham (thin)	4 eggs (poached)
1 tablespoon butter	Hollandaise Sauce (p. 283)
2 English muffins (cut in half and toasted)	truffles (optional)

Sauté the ham slices in sizzling butter for a few seconds on each side. Place ham slices on top of toasted English muffins. Meanwhile, poach eggs and make Hollandaise sauce. Poach eggs by adding salt and white vinegar to boiling water, then reduce to a simmer and introduce the eggs gradually—breaking each onto a saucer and sliding it into the water. When the whites are firm, remove eggs with a slotted spoon and drain them on paper towels. Place the eggs on top of the ham and muffin. Cover them with Hollandaise Sauce. The classic embellishment is a slice or two of truffles. Of course these have become so expensive that even fine restaurants often do not use them.

⸗ *Serve with asparagus, which will go nicely with any of the Hollandaise Sauce that seeps over from the eggs, and a dry white wine.*

FRANKFURTER AND BAKED BEAN CASSEROLE

Serves: 4
Time: 30 minutes

2 13-ounce cans tiny baked beans	2 tablespoons ketchup
	2 tablespoons wine vinegar
2 tablespoons chopped onions	8 large frankfurters

Preheat oven (325° F.). Place baked beans in casserole. Add onions, ketchup, and vinegar. Stir well. Meanwhile, slit frankfurters in about 4 places and place on top of beans. Cover casserole and cook for 30 minutes.

⸱ *Tip: If beans dry out a bit, just add a dash of red wine to give them extra moisture.*
⸱ *Serve with any vegetable in the cabbage family—broccoli, Brussels sprouts, etc.—but this is a real meal in itself, so why not serve just a good coleslaw (with a little mustard added)? Beer would be a good drinking accompaniment.*

8

Poultry

TO SOME PEOPLE, poultry means chicken, but the poultry category also includes such year-round staples as turkey, duck, and goose—as well as such delicacies as guinea and Rock Cornish game hens. Wild birds also fit into the poultry group—pheasant, grouse, quail, and so on (these are discussed in Chapter 10).

⌐ WHEN YOU BUY POULTRY

The U.S. Government inspects poultry as it does meat. Grading for wholesomeness is compulsory in any interstate business. Government inspectors and veterinarians constantly inspect poultry farms and plants for cleanliness and they check the entrails of every bird to make sure it is not diseased. The USDA stamp should appear on every bird. It is a round seal that looks like this:

The government grading stamp for quality is optional and is provided to poultry-processors who pay a fee. The grades are

A, B, and C, and they correspond to prime, choice, and good in meat. The stamps are shieldlike and certify that the poultry has been graded for quality by a technically trained government grader. When the grading is done in cooperation with a state, the official grade stamp may include the words Federal-State Graded.

U.S. Grade A designates the highest quality. It means that the birds are fully fleshed and meaty, that they are attractive in appearance and "well finished," and that they have been fed quality grains. The B- and C-graded birds are not as plump or attractive. They may even be a bit scrawny, with the breast bone jutting up and not well covered with meat. Few poultry farmers bother to have the B and C grades designated. A is their best product and that is what you will find in the finest butcher shops—along with prime-quality meat.

The grade of the poultry you buy does not really indicate the tenderness of the bird—it merely tells you its quality. For example, young fowl and mature birds may both be graded A because of their excellent conformation and quality. But their ages will require different cooking methods. Frequently the wholesomeness stamp and the grade stamp are included in one overall shield that also tells the type of each bird (frying chicken, stewing chicken, etc.).

Your butcher can give you accurate information about age, quality, and so on, but for supermarket identification, it is well to know the labeling for young and mature fowl.

YOUNG TENDER-MEATED FOWL

These, of course, are most suitable for frying, broiling, barbecuing, and even roasting.

YOUNG CHICKENS
These may be labeled: young chicken, broiler, fryer, roaster, capon, or Rock Cornish game hen.

YOUNG TURKEYS
May be labeled: young turkey, young hen, young tom, or just turkey.

YOUNG DUCKS
The labeling may be: duckling, young duckling, broiler duckling, or roaster duckling.

MATURE FOWL

These are less tender and need more cooking. They are excellent for stewing, soups, long roasting, and salads.

FACTS ABOUT CHICKEN

Chicken is generally recognized as a high-quality protein food that is quite low in calories. It has high vitamin content and low fat content (whatever fat there is seeps off during cooking). A 3-ounce serving of broiled chicken meat will provide 29 percent of your daily protein requirement.

As mentioned previously, the tenderness of a bird depends

on its age, and the age relates directly to its poundage. For example, a broiler-fryer is usually about eight to nine weeks old and weighs from 1½ to 4 pounds. A roaster is about fourteen weeks and weighs a bit more. Stewing chickens reach a much older age, sometimes as much as ten months, and weigh up to 7 pounds. These are slaughtered when the farmer decides they have become less important as breeders or egg-layers.

The best way to tell a good chicken is to look for a rich yellow coloring. It should be plump in shape, which means that it has firm meat and a good inner coating of fat. Forget a chicken that looks scrawny or dried-up.

CAPON

These delicious roasting birds usually have to be ordered in advance, even though your butcher considers them his specialty. Our pièce de résistance (which is an all-time party favorite) is boned capon stuffed with another boned capon. A capon is a male bird from which the reproductive organs have been removed. The surgery is done when the bird is only four weeks old. These birds do not crow and show no desire to fight. The growth of comb and wattles do not keep pace with the rest of the body, which gives his head a long and undeveloped appearance. Because of its tranquil life, the capon's flesh retains the fine flavor and texture of broiler meat. He is larger and plumper than a regular chicken roaster and has an exceptionally good flavor. Capons usually age to about seven months.

ROCK CORNISH GAME HENS

These delectable little birds were actually "invented." It took a printer of exceptional engravings—with a taste for fine art and fine food—to dream up the idea. Jacques Makowsky and his wife, Therese, came to America via Russia and Frence. He practiced his art in New York City for eight years and then retired to a farm in Pomfret Center, Connecticut. The wild beauty of his surroundings—plus his idleness—caused him to call his farm Idle Wild. He decided to become a poultry farmer, but only of gourmet birds. He first raised Guineas, but later he and his wife, after much experimenting, came up with a crossbreed of Cornish game cocks and Plymouth Rock hens. The result was a plump little bird with all-white meat.

The first Rock Cornish game hen made its debut in 1950. Now they sell in the millions and are shipped all over the country to gourmet restaurants and fine butcher shops. Aside from the succulence of its white meat, it has a distinct gamy flavor. This is because the birds are given a high protein diet that includes such native Connecticut produce as cranberries, acorns, and other nuts. It is sometimes compared in appearance and taste with the less available quail and squab.

The usual weight is 1 pound, which is ample for an individual serving. In fact, half a bird is adequate for home parties —if the rest of the menu is abundant.

FACTS ABOUT GUINEA HEN

The guinea fowl was a wild bird when it was first introduced to Europe in the sixteenth century—from Africa by way of Turkey. Most guineas have feathers that are dark gray (with a somewhat lavender tinge), with dots of a pearl-white color. But there are also white African guineas. Their heads and necks are usually bare and have a scrawny look, but sometimes they have a rather interesting feathered crest. Any type of guinea fowl has a slight and pleasant gamy taste. The meat is usually white but slightly drier than that of other birds (so don't overcook!). Should you decide to have a fancy guinea dinner, remember that the hen is much more tender than the cock.

FACTS ABOUT TURKEY

Evidence indicates that early Indian tribes (Aztec, Maya, Inca) had domesticated turkeys long before Columbus made his appearance, and the tradition of serving turkey on the Thanksgiving board started way back in 1621.

Ben Franklin was so admiring of the turkey that he wanted it—instead of the bald eagle—to appear on the Great Seal of the

United States. In a letter to his daughter, he wrote: "I wish the bald eagle had not been chosen as the representative of our country; he is a bird of bad moral character; like those among men who live by sharping and robbing, he is generally poor and often lousy . . . the turkey is a much more respectable bird, and withal a true original native of America."

Although turkey was originally a specialty for holiday tables, it has now become an all-season favorite, with Americans eating nearly twice as much of it as they did just a short time ago. Modern production and processing methods have made turkey available year-round. And they are plumper, meatier, and more compact, with a larger proportion of breast meat. They have a low fat content and are high in B vitamins. Turkey also heads the list of lean meats in protein content. A study at Cornell University established turkey as highest in protein and lowest in cholesterol of all other poultry, and of all red meat except veal.

TURKEY-BUYING TIPS

Unlike the golden yellow color of chicken, turkeys are usually white, even faintly blue. They should be plump and rounded over the breast bone.

For a ready-to-cook (or dressed) turkey of 12 pounds or less, figure ¾ to 1 pound for each guest. But if the bird is over 12 pounds, figure ½ to ¾ of a pound for each guest.

Turkeys are usually raised for from eighteen to twenty-one weeks. The hen weighs from 5 to 16 pounds and the tom from 16 to 32. Even with weight variations, the hen and the tom of the same age are almost equal in tenderness, but the hen is somewhat more tender.

The younger they are, the more tender. Tenderness is also influenced by cooking methods. The slow process is always best with turkey. Although many people prefer to buy a fresh bird, about 90 percent of all whole carcass turkeys are retailed "fresh-frozen." They are also available in frozen sections, a convenience for small families.

FACTS ABOUT DUCK

Long Island is the greatest producer of ducks in the United States. Almost 8 million ducklings are produced each year on various Long Island farms, and in the plants of the Long Island Duck Farmers Cooperative, Inc., 20,000 ducklings can be flash-frozen each day at 60° below zero. Then they are shipped to every state, as well as to Europe.

Long Island ducklings are descended from the white pekin ducks that started their journey to America in 1873 from China. The captain of a yankee clipper accumulated a small flock in China. Only a few survived the long trip around Cape Horn, but some of them were placed on a Long Island farm, where they took to the climate, the sandy shores, and the abundant water supplies and produced and reproduced.

Advanced feeding methods and extreme care and cleanliness have developed prime birds. In fact they are so admired that the original source of white pekin ducks—the Orient—now imports Long Island duckling.

DUCK-BUYING TIPS

If you have seen Long Island ducklings in their home habitat, you will know how beautiful and white their feathers are. When you buy a bird at your butcher shop, you will also find that the skin is almost white (not yellowish like a chicken's). These plump, tender birds are usually only about seven or eight weeks old and weigh from 4½ to 5 pounds. Of this weight there is a considerable amount of fat, which should be drained off during cooking so that the skin will be crisp and golden.

A duckling is adequate for four people. It is so young and tender that it can easily be quartered with a sharp knife or poultry shears. Each quarter is an ample serving, especially when stuffing is involved. But do give your special guests the breast sections, as they are meatier.

Restaurants usually serve half a duckling, which is a bit too much for a single serving, except for very hearty eaters.

FACTS ABOUT GOOSE

Like turkey, goose used to be a holiday specialty, but it too is now available year-round because of quick-freezing. This bird is known for its high fat content, but growers are constantly working on feeding formulas to produce a meatier bird —so you won't have to drain off so much fat while it is cooking.

Years ago, goose was considered a somewhat peasant-type dish. But now it has gained gourmet status and is promoted by restaurants who specialize in elegant and unusual food. The goose is extremely sweet-meated, tender, and juicy, and it is the answer for those who crave dark meat—all of the goose is just that.

GOOSE BUYING TIPS

As with most birds, youth is an important factor where tenderness is concerned. However, geese are available in varying weights—from 6 to 20 pounds—and the weight, of course, indicates the youthfulness. In buying, always judge the ready-to-cook weight. And because of the fat loss in cooking, it is safe to allow 1 pound of goose for each individual serving.

BARBECUED CHICKEN NO. 1

Serves: 6 to 8
Time: Barbecue sauce, 35 minutes;
chicken, 30 to 45 minutes

2 broilers (quartered)	1 tablespoon Worcestershire
2 8-ounce cans tomato sauce	sauce
2 teaspoons sugar	¼ cup bouillon
salt to taste	1 tablespoon cognac

Wash chicken and dry with paper towels. Combine the rest of the ingredients (except cognac) and simmer covered for about 20 minutes. Remove from heat and immerse chicken pieces in the sauce until they are well covered. Place chicken on a platter, cover with aluminum foil, and refrigerate overnight. Continue

cooking sauce for 15 minutes. Cool, cover, and store in refrigerator overnight. Before barbecuing the next day, bring chicken
and sauce to room temperature and build a charcoal fire. When
coals have an even glow, place chicken quarters on grill, skin
side up. Baste with sauce and turn about every 10 minutes.
Depending on the heat of the coals, cook for 30 to 45 minutes.
Just before you serve chicken, heat the remaining sauce and
add the cognac. Use this as a dip or pour it over the chicken.

⟡ *Serve with baked macaroni with cheese, Italian string beans, and
a medium-dry red wine.*

BARBECUED CHICKEN NO. 2

Serves: 8
Time: About 45 minutes

2 3-pound fryers (quartered)
½ cup Chinese sauce (Hoisin)
2 tablespoons oil
½ cup sherry

salt and freshly ground pepper
2 cloves garlic (minced)
3 scallions (chopped)

Wash and dry chicken. Brush well with Chinese sauce and
let stand until it has been absorbed. Meanwhile, mix the rest
of ingredients in a large bowl. Place the chicken in this marinade and let stand at room temperature for 3 hours or more.
When the charcoal has settled down to red coals at the bottom
and gray on top, place the chicken pieces on the grill (skin
side up). Turn the chicken and baste with the marinade about
every 10 minutes. The complete cooking time should be about
45 minutes, depending on the strength of the heat. Put the last
bit of marinade over chicken just before you serve.

⟡ *Serve with fluffy rice, broccoli, and a cool rosé wine or beer.*

SWEET AND SOUR BARBECUED CHICKEN

Shelley Winters is an exceptionally good cook and knows many food tricks. She is particularly fond of cooking on her prized hibachi.

Serves: 2
Time: About 50 minutes

1 2-pound chicken (halved)	salt and freshly ground pepper
1 clove garlic	spicy sauce (Beau Monde)
1 cup honey	water
½ cup soy sauce	

The first thing Miss Winters does is to pound the chicken halves with a hammer. Then she rubs the chicken with a peeled clove of garlic. Meanwhile, she mixes the honey, soy sauce, and salt and pepper. She adds a dash of spicy sauce and blends the mixture together with enough water to give it a syruplike texture. This marinade is supposed to cover the chicken for at least 3 hours. When placed on the hibachi, it should be cooked over a steady (but low) flame for 20 minutes. Then turn, baste it with the marinade, and cook 20 minutes more; finally turn, baste, and cook another 5 minutes on each side.

⸕ *Miss Winters serves wild rice with this (sprinkled with parsley), arugula salad with safflower oil, and Spanish white wine (not too dry).*

BROILED CHICKEN HALVES

Serves: 4
Time: 40 to 50 minutes

2 broilers (2 to 2½ pounds each)	2 cloves garlic (crushed)
	salt and freshly ground pepper
½ cup oil	pinch of thyme
¼ cup white wine	

Ask butcher to cut broilers in half and remove back and

breast bones. Wash chicken halves and dry well. Preheat oven (350° F.). Now combine the oil, wine, garlic, salt and pepper, and thyme. Brush this mixture over chicken halves and place them in shallow baking pan, skin side up, and bake for about 20 minutes. Remove from oven and turn heat up to broil. Meanwhile, brush marinade over chicken on both sides. Return to broiler, skin side down, and broil for 10 minutes. Turn chickens skin side up, brush with marinade, and broil for another 10 minutes or until skin is crisp.

ǀ Serve with fluffy rice (pour sauce over it), green peas, and a medium white wine.

CHICKEN BREASTS WITH ORANGE SAUCE

Serves: 4 to 6
Time: About 1 hour

4 chicken breasts (cut in half)
salt and freshly ground pepper
3 tablespoons butter
1 clove garlic (minced)
¼ cup flour

½ teaspoon cinnamon
1 teaspoon parsley (minced)
2 cups orange juice
1 small can mandarin oranges

First of all, wash chicken breasts and dry well with paper towels. Sprinkle them on all sides with salt and pepper and bring to room temperature. Melt the butter in a large skillet and add garlic. Before it has become brown, add the chicken pieces and brown on all sides. Remove chicken to drain on paper towels. To the butter remaining in skillet, gradually add flour, cinnamon, and parsley. Stir constantly until thickened, then slowly add the orange juice. Continue stirring until sauce comes to a boil. Reduce heat and add the browned chicken breasts. Cover and simmer for about 50 minutes. Now add the mandarin oranges and cook for another 5 minutes.

ǀ Serve with glazed carrots, buttered peas, and a very dry chilled white wine.

BRAISED CHICKEN WITH RAISINS

Serves: 4
Time: About 1¼ hours

1 3-pound frying chicken	4 tablespoons butter
2 tablespoons flour	1½ cups orange juice
salt	½ cup seedless white raisins
½ teaspoon cinnamon	

Have chicken cut in eighths. Wash chicken pieces and dry thoroughly. Mix flour, salt, and cinnamon well and roll chicken pieces in this. Heat butter in large skillet and brown chicken on all sides. Add orange juice and bring to a boil. Reduce heat, cover skillet, and simmer for about 1 hour. Should sauce dry out, add a bit more orange juice. Twenty minutes before end of cooking time, add the raisins.

⁄ *Serve with fluffy rice, Brussels sprouts with lemon butter, and a chilled Chablis wine.*

OVEN-FRIED CHICKEN NO. 1

Serves: 4
Time: About 1 hour

1 large frying chicken (about 3 pounds)	1 cup potato chips (crushed)
½ cup oil	1 cup corn flakes (crushed)
salt and pepper	parsley sprigs

Have butcher cut fryer in eighths. Wash chicken pieces and dry them well. Preheat oven (350° F.). Brush chicken pieces with oil, sprinkle with salt and pepper, and then roll them in the combined potato chips and corn flakes until they are completely coated. Arrange chicken pieces in a single layer on a large shallow baking pan. Sprinkle them with a little oil and bake in oven for about 1 hour.

⁄ *Serve with corn on the cob, chopped spinach, and a cool rosé wine.*

OVEN-FRIED CHICKEN NO. 2

Serves: 4
Time: About 1 hour

1 large frying chicken (about
   3 pounds)
1 lemon (quartered)
½ cup flour
salt and freshly ground pepper

2 eggs (beaten)
¾ cup seasoned breadcrumbs
¼ pound butter
paprika

Have butcher cut chicken into pieces. Rub lemon quarters over chicken on all sides and let stand for 10 minutes or more. Preheat oven (375° F.). Then dry chicken and cover with flour flavored with salt and pepper. Dip each piece of chicken into the beaten eggs and then roll them in the breadcrumbs. Melt the butter in a large shallow baking pan and add the chicken. Be sure pan is large enough so that chicken pieces do not touch. When pieces are lightly browned on one side, turn them and bake for about 1 hour. Turn them again midway through baking. They should be brown enough—but if they are not, turn up the oven a bit. Sprinkle generously with paprika.

⸢ *Serve with buttered Fordhook lima beans, a little salad, and a touch of light white wine.*

CRISPY OVEN-FRIED CHICKEN BREASTS

Serves: 4
Time: 1¼ hours

chicken breasts from 2 large
   fryers
1½ cups corn flakes (crumbled)
salt and freshly ground pepper

¼ teaspoon garlic powder
¼ teaspoon paprika
2 eggs (beaten)
½ cup cooking oil

Have butcher bone breasts and cut them in half. Wash them and dry well. Preheat oven (350° F.). Combine corn-flake crumbs with salt and pepper, garlic powder, and paprika. Dip the breast pieces in the egg and then roll them in the seasoned corn-flake crumbs. Line a shallow baking dish with aluminum foil and place the chicken breasts on foil and bake for 10 minutes. Meanwhile, heat the oil in a frying pan. When very hot, drizzle the oil over chicken. Bake uncovered for about 1 hour.

/ *Serve with paprika potatoes, buttered asparagus, and a chilled white wine.*

DEEP-FRIED CHICKEN IN BATTER

Serves: 3 (maybe more)
Time: 10 to 15 minutes

1 2½-pound fryer	1 cup flour
oil	1 cup beer
1 cup prepared biscuit mix	parsley sprigs

Have butcher cut chicken in eighths. Wash each piece and dry well. Preheat oil in deep fryer to 450° to 475° F. Combine biscuit mix, flour, and beer until it becomes a gluey mixture. Add a little water if needed. Use tongs to dip each piece of chicken into the batter and then place in basket of deep fryer. When crisp and well browned, place pieces on paper towels to drain. Serve on a large hot platter and decorate with parsley sprigs.

/ *As the crust on the chicken provides enough starch, just serve with buttered string beans, sliced tomatoes with French dressing, and ice-cold beer.*

BAKED CHICKEN HALVES WITH WINE

Serves: 2
Time: About 45 minutes

1 2-pound broiler	paprika
1 lemon	1 cup dry white wine
2 tablespoons butter	thyme
salt and freshly ground pepper	rosemary

Ask butcher to cut broiler in half. Cut lemon in quarters and squeeze juice all over inside of chicken and let stand for 10 minutes or more. Preheat oven (450° F.). Turn and squeeze juice on skin side, rubbing pulp against skin, and let stand as long as possible. Heat butter in flat baking pan. Place chicken in pan, skin side up, and sprinkle with salt and pepper and paprika. Cook in hot oven for 5 minutes. Then add wine and pinch of thyme and rosemary. Return to oven for 5 more minutes. Now reduce heat to 325° F. and cook for 35 minutes, basting and adding wine if needed.

⁊ *Serve with riced potatoes and asparagus—and a nice dry white wine—a Chablis, for example.*

CHICKEN JUBILEE CASSEROLE (FLAMBÉ)

Serves: 4
Time: About 1½ hours

1 3½-pound fryer
1 cup French dressing
1 can frozen orange-juice concentrate
1 cup pitted black cherries
½ cup currant jelly
2 tablespoons brandy

Have butcher cut chicken in eighths. Place chicken in a large bowl and marinate with French dressing for 1 hour, turning it once or twice so that marinade will reach every section of the chicken. Preheat oven (400° F.). Melt orange juice in a saucepan and add cherries and jelly. When this mixture is well heated, drain the chicken and place in a casserole, then pour the hot sauce over it. Cover the casserole and bake in oven for about an hour. If sauce seems too watery, uncover casserole for last 30 minutes of cooking. Just before serving, heat the brandy in a ladle, flame it, and pour over chicken dish.

⁊ *Serve with mashed sweet potatoes, broccoli, and a chilled Sauterne wine.*

CHICKEN CASSEROLE FLAMBÉ

Serves: 4 to 6
Time: About 1¼ hours

1 3½-pound chicken	pinch of thyme
4 tablespoons butter	salt and freshly ground pepper
1 large onion (chopped)	½ cup bouillon
1 large celery stalk (chopped)	2 teaspoons cornstarch
¼ green pepper (chopped)	1 cup heavy sweet cream
2 carrots (diced)	2 tablespoons brandy
½ teaspoon rosemary	

Have chicken cut in eighths. Wash and dry each piece well. Place butter in a large skillet and let it bubble. Add the onion, celery, pepper, and carrots and cook them for about 10 minutes or until they are lightly browned. Stir in the rosemary, thyme, and salt and pepper, then add the bouillon and let it come to a boil. Reduce heat and add the chicken pieces. Cover and simmer for about 45 minutes. Remove from heat and allow to cool for 15 to 20 minutes. Preheat oven (300° F.). Now remove the chicken pieces and place in a casserole. Strain the sauce into a blender (discarding vegetables). Add cornstarch and cream to blender and blend for about 30 seconds. Pour this mixture over the chicken. Cover and bake for about 15 minutes. Before serving, heat the brandy in a ladle, flame, and pour over chicken.

/ *Serve with creamy mashed potatoes, green peas, and a chilled Sauterne wine.*

CHICKEN AND WINE CASSEROLE WITH APPLES

Serves: 4
Time: About 1¼ hours

1 3-pound fryer	2 cloves garlic (minced)
salt and freshly ground pepper	2 large apples (sliced)
4 tablespoons butter	1 cup dry white wine

Have butcher cut chicken in eighths. Wash pieces and dry well. Season with salt and pepper and bring to room temperature. Preheat oven (350° F.). Melt butter in large skillet. Add garlic and stir for a minute, then add chicken pieces and brown lightly on all sides. Place browned chicken in bottom of large casserole with apple slices on top. Pour wine over all. Cover tightly and bake for 1¼ hours.

☙ *Serve with candied yams, buttered broccoli, and a chilled dry white wine.*

CHICKEN LEGS WITH SOY SAUCE

Serves: 4
Time: About 35 to 40 minutes

4 chicken legs
½ lemon

2 tablespoons soy sauce
2 tablespoons dry vermouth

Have butcher disjoint legs at the thigh. Rub leg pieces well with lemon juice and place in a shallow baking pan. Sprinkle 1 tablespoon of the soy sauce over legs and let stand for 15 minutes or more. Preheat oven (350° F.). Turn the legs and sprinkle the other tablespoon of soy over them. After they have stood at room temperature for another 15 minutes, place pan in oven with skin side up. You will find that the soy sauce will start to dry out after 10 to 15 minutes. This is the time to loosen it up with the vermouth. But keep watching and baste with the pan juices. You may have to add a bit more vermouth. During the last 10 minutes of cooking reduce the heat to 300° F.

☙ *Rice, of course, is again the great accompaniment. Since soy sauce is involved, how about some Chinese cabbage? Any chilled dry white wine will go well.*

CREAMED CHICKEN—SOUTHERN STYLE

Serves: 6 to 8
Time: About 1¼ hours

2 2½-pound broilers	2 tablespoons flour
salt and freshly ground pepper	1 cup chicken broth
2 tablespoons chicken fat	1 cup light cream
1 large onion (diced)	paprika
1 cup mushrooms (sliced)	

Ask butcher to quarter chickens. When you get them home, wash them and dry with paper towels. Then sprinkle them with salt and pepper and bring to room temperature. Preheat oven (325° F.). Melt the chicken fat in a large skillet and sauté the onions and mushrooms. Push them to one side of skillet and brown the chicken quarters on all sides, adding more fat if needed. Place chicken in a flat baking dish and reserve mushrooms and onions. Add flour gradually to the fat left in frying pan, stirring constantly, then add chicken broth a little at a time until it is bubbling. Reduce heat and stir in the cream. When it is well heated, add this sauce to the chicken. Cover and bake for about 30 minutes. Add the sautéed mushrooms and onions to the cream sauce and stir well. Return to oven with the cover on and bake for another 20 minutes. If sauce has diminished, add a bit more broth or cream. Before serving, sprinkle with paprika.

⚹ *Serve with brown rice and baby lima beans—a cool rosé wine would be nice too.*

CHICKEN ALABAMA STYLE
WITH CREAM SAUCE

Serves: 6 to 8
Time: About 1 hour

2 2½-pound fryers
6 strips crisp bacon (crumbled)
salt and freshly ground pepper
¼ teaspoon garlic powder
paprika (about ½ teaspoon)
1 cup flour

4 tablespoons oil
6 tablespoons butter
2 tablespoons flour
1 cup light cream
3 tablespoons parsley (chopped)

Ask butcher to cut fryers in eighths. When you get them home, wash them and pat them dry with paper towels and bring to room temperature. Preheat oven (350° F.). In a large bowl, combine the crumbled bacon, salt and pepper, garlic powder, paprika, and flour. Dip each piece of chicken in this mixture until it is well covered. Place pieces on wax paper after dipping. Heat oil in a very large skillet and then add 4 tablespoons of the butter. When sizzling, add the chicken pieces and brown on all sides. Do not crowd the skillet; cook only a few pieces at a time. As each piece becomes golden-brown, place it in a shallow baking pan. Bake chicken for half an hour. Remove chicken to a hot platter. Melt the rest of the butter in the same frying pan in which the chicken was fried. Gradually stir in the flour, then the cream. When it is hot but not boiling, pour over chicken and sprinkle with parsley.

✟ *Serve with turnip greens, fluffy rice, and a dry white wine.*

CHICKEN SESAME

Virginia Graham, whose nationwide television show has captured audiences from coast to coast, is also a lecturer and an author, and she was the first woman to be National Crusade Chairman of the American Cancer Society. In this post she was their most successful fund-raiser ever and is now their Honorary Chairman of Education. The recipe she gave us is one of her favorite ways of preparing chicken.

Serves: 12
Time: About 40 minutes

6 small frying chickens
  (halved)
4 eggs (beaten)
½ cup heavy sweet cream
2 cups flour
2 teaspoons baking powder
2 teaspoons salt

3 teaspoons paprika
¾ teaspoon freshly ground
  pepper
1 cup cashew nuts (chopped)
½ cup sesame seeds
1 cup butter

Rinse chicken in cold water and wipe dry. Combine eggs and cream and dip the chicken into this mixture. Preheat oven (400° F.). Meanwhile, place all of the other ingredients (except butter) into a paper bag. Place the chicken halves, one at a time, into this mixture and shake well so that all of the ingredients adhere to the chicken. Then place the butter in a shallow baking pan and heat over top of stove. When sizzling, place chicken halves in butter. Coat them and turn them so that all pieces are buttered on all sides. Now place pan in oven with skin side down. Bake for 20 minutes. Turn chicken and bake for another 20 minutes. If the guests are not quite ready for dinner, reduce the heat and brush with melted butter.

ı Although Miss Graham did not mention accompaniments, we suggest baby peas with tiny onions and a chilled dry white wine.

STEWED CHICKEN WITH DUMPLINGS

Serves: 6
Time: About 3 hours

This is a dish that is known from Maryland down through all of the southeastern states. It is a meal in itself and needs only a bit of greenery to accompany it.

1 6- to 7-pound stewing chicken	2 cups flour
4 carrots (chopped)	1 tablespoon baking powder
2 onions (quartered)	1 teaspoon salt
salt and freshly ground pepper	⅓ cup shortening
1 bay leaf	½ to ⅔ cup milk

Place chicken, carrots, onions, salt and pepper, and bay leaf in a large kettle or Dutch oven. Add water to cover and bring to a boil. Reduce heat and simmer for about 2 hours, covered. Now remove chicken and discard bay leaf. When chicken is cool enough to handle, remove meat from bones. Discard bones and cut meat into strips; return to broth. Meanwhile, cool ½ cup of the broth and stir in 2 tablespoons of the flour. Add this mixture to the broth so that it will thicken and let simmer. Now sift the rest of the flour with the baking powder and salt. Cut in the shortening until mixture is crumbly. Gradually add the milk to make a stiff dough. Turn this mixture out onto a lightly floured board or pastry cloth and knead gently for about 30 seconds. Then roll it out to about a ⅛-inch thickness. Cut into ½" by 4" strips. Drop these strips on top of the bubbling broth and chicken. Cover tightly and simmer gently for about 30 minutes.

✓ *Serve with a crisp green vegetable or a mixed green salad and a chilled dry white wine.*

CHICKEN FRICASSEE

Serves: 4 to 6
Time: About 1½ hours

1 4-pound chicken (cut in eighths)
salt and freshly ground pepper
paprika (about 1 teaspoon)
¾ cup flour
2 tablespoons chicken fat
1 large onion (diced)
2 cloves garlic (minced)
1 bay leaf
2 cups stewed tomatoes
1 green pepper (chopped)
4 medium potatoes (peeled and quartered)

Wash chicken and pat dry with paper towels. Mix salt and pepper and paprika with flour and roll chicken pieces in this mixture. Heat the chicken fat in a large deep skillet and add onion, garlic, and bay leaf. When onion is transparent, move to one side and brown the chicken on all sides. Add the stewed tomatoes and peppers and, when sauce comes to a boil, reduce heat and simmer covered for about 45 minutes. Add the potatoes and continue simmering for about 30 minutes. Discard bay leaf.

/ *Inasmuch as potatoes are involved in the fricassee, there is no need for a starchy vegetable. So just pick your favorite green one and have sliced tomatoes with French dressing.*

CHICKEN IN WHITE WINE

Serves: 6 to 8
Time: About 1¼ hours

2 2½-pound chickens
⅓ cup corn oil
½ cup salt pork (diced)
½ cup scallions (chopped)
24 small white onions
⅓ cup cognac (warmed)
2 teaspoons garlic (chopped)
1½ teaspoons salt
freshly ground pepper
2 cups Chablis wine
1 pound medium mushrooms

Have chicken cut in serving pieces. Wash and dry thoroughly. Place corn oil in a large heavy pot and brown the chicken on all sides, a few pieces at a time. Remove each piece as it is browned and set aside. Now add the salt pork, scallions, and white onions. Brown lightly and then spoon off the fat. Replace the chicken. Set the cognac alight and pour flaming over the chicken. Place the garlic and salt on a board and crush them together with a knife blade until smooth. Add this mixture to the chicken with plenty of pepper and the wine. Cover and simmer for 20 minutes. Add the mushrooms and continue simmering, covered, for 30 to 40 minutes.

COQ AU VIN—THE EASY WAY

Serves: 4 to 6
Time: About 1¼ hours

This is a classic dish, although individual chefs have their own versions. Here is an uncomplicated, easy-to-prepare version.

1 4-pound frying chicken	2 cups sherry
2 tablespoons butter	1 bay leaf
3 large onions (quartered)	1 tablespoon parsley (chopped)
1 clove garlic (crushed)	thyme (about ¼ teaspoon)
6 medium carrots (scraped)	½ cup mushrooms (chopped)
2 tablespoons flour	salt and freshly ground pepper

Have butcher cut chicken in eighths. Wash and dry the chicken thoroughly. Melt the butter in a large frying pan and brown chicken pieces on all sides. Remove to paper towels to drain. In the same skillet, add onions, garlic, and carrots. Let these brown slightly. Push them to one side and sprinkle in the flour, stirring constantly, then gradually add the sherry. When well mixed, add bay leaf, parsley, thyme, and mushrooms. Sprinkle this mixture with salt and pepper. Reduce heat and add the chicken. Allow to simmer for about 45 minutes, covered. Add more wine if needed. Discard bay leaf.

⟩ *Serve with baby French peas and tiny onions—and, of course, champagne.*

CURRIED CHICKEN

Serves: 6
Time: About 1¾ hours

Curry dishes are especially popular for parties—because like some other casseroles they can "wait" a bit while the company decides when they wish to eat. You really have to know your guests to decide exactly how much curry to use. This recipe has a medium amount of curry, so you will have to try it out to see if you wish to use more or less.

1 5-pound stewing chicken	2 tablespoons curry powder
1 teaspoon salt (approximately)	2 cups chicken stock
3 tablespoons chicken fat	1 tablespoon flour
1 large onion (diced)	1 egg yolk (beaten)

Have butcher cut chicken into serving pieces. Place these pieces into large pot with just enough water to cover chicken. Bring to a boil and add salt. Cover pan, reduce heat, and simmer for about 1 hour. Remove chicken and drain (reserve stock). Melt the chicken fat in a large pan and add onions. When they are lightly brown, move to one side and brown the chicken on all sides. Mix half of the chicken stock with the curry powder and pour into the chicken and onion pot. Let this simmer for about 5 minutes. Now take a tablespoon of cooled chicken stock and add flour and egg yolk. Add this mixture to the pan and stir constantly. As it thickens, add more of the stock.

✦ *Serve with long-grain fluffy rice. But the most important thing with curry dishes are the asides—shredded coconut, chopped nuts, raisins soaked in brandy, and, of course, chutney.*

CHICKEN AND TOMATO STEW
WITH CURRY

Serves: 6 to 8
Time: About 1¼ hours

2 2½-pound broilers
¼ pound butter
3 medium onions (chopped)
1½ cups green peppers
  (chopped)
4 cloves garlic (minced)
½ cup flour

salt and freshly ground pepper
2 tablespoons curry powder
½ teaspoon thyme
½ cup lemon juice
2 8-ounce cans tomato sauce
½ cup water

Have butcher cut chickens in eighths. Wash and dry each piece. Place butter in a very large skillet and, when sizzling, add the onions, peppers, and garlic. When they are limp but not brown, remove to a side plate. Meanwhile, dip chicken pieces in flour seasoned with salt and pepper. Place chicken in bubbling butter in skillet and brown on all sides, then remove to paper towels. In the remaining butter in skillet, stir in the curry powder, thyme, and 4 tablespoons of the flour. If the butter has diminished, add a bit more so that the flour will blend into a paste. Gradually add to this paste the lemon juice, tomato sauce, and water. Stir until smooth. When it is bubbling, reduce heat and return the cooked onions, pepper, garlic, and the chicken pieces. Cover the skillet and simmer for about 1 hour. Should sauce thicken too much, add a bit more water.

/ *Serve with rice cooked with raisins. Also, chopped almonds, chutney, and shredded coconut. Then have a large bowl of salad and either a white wine or beer.*

CHICKEN GUMBO—WITH A NEW ORLEANS FLAVOR

Serves: 4 or more
Time: About 1½ hours

1 4-pound chicken	1 bay leaf
2 celery stalks (chopped)	1 cup stewed tomatoes
salt and freshly ground pepper	1 cup okra (canned or fresh)
1 large onion (chopped)	¼ cup uncooked rice
3 tablespoons butter	1 tablespoon parsley (chopped)
1 green pepper (chopped)	

Place chicken in a large pot with enough water to cover. Add celery, salt and pepper, and onion. Cover and cook for 1 hour. Remove chicken and let cool, reserving the broth. Bone the chicken and dice the meat. Meanwhile, melt the butter in a large frying pan and add the chopped pepper. Let the pepper simmer and return the diced chicken to 2 cups of the chicken broth along with the bay leaf and cook over low heat. When the peppers have softened, add them and the butter to the chicken—also add the tomatoes, okra, rice, and parsley. Simmer this mixture for about 30 minutes. Discard bay leaf.

/ This is a very juicy dish that includes rice, so no other starchy vegetable is needed. Why not just settle for crisply toasted garlic bread and a wonderful salad? A medium-dry red wine would also be excellent.

CHICKEN CACCIATORE

Serves: 4
Time: About 1¾ hours

1 4-pound broiler	1 8-ounce can tomato sauce
½ cup flour	2 medium tomatoes (quartered)
4 tablespoons oil	½ cup dry white wine
1 large onion (chopped)	salt and freshly ground pepper
3 cloves garlic (minced)	pinch of thyme
½ green pepper (diced)	

Have chicken cut in eighths. Wash and dry chicken and dredge with flour. Heat oil in large skillet and add onion, garlic, and green pepper. When onion is transparent and pepper limp, push to one side of skillet and brown the chicken pieces. Remove browned chicken to a Dutch oven. To the skillet, add tomato sauce, tomatoes, wine, salt and pepper, and thyme. When this mixture is bubbling, pour it over the chicken. Place covered Dutch oven over low heat and simmer for about 1¼ hours. After the first half-hour, stir the mixture. If it seems to have dried out, add a bit more wine.

↲ *Serve with large noodles and a mixed green salad. A medium-dry red wine tastes good too.*

CHICKEN ALMOND SOUP

Serves: 6
Time: About 30 minutes

1 cup cooked chicken
1½ tablespoons butter
1½ tablespoons cornstarch
4 cups chicken broth
1½ cups heavy sweet cream

½ cup almonds (finely chopped)
sprinkle of nutmeg
1½ tablespoons chives (chopped)

Shred or chop the chicken. In a large pot, melt the butter and gradually add the cornstarch. When this mixture has become smooth and pasty in texture, add the chicken broth slowly and stir constantly. Bring to a boil, reduce heat, and add the chicken. Allow to simmer for about 15 minutes, then add the cream, almonds, and nutmeg. Again simmer (but do not boil) for about 15 minutes. Serve in hot soup dishes and garnish with chopped chives.

↲ *This is a delightful luncheon dish and can be served hot or chilled. All you need with it is an excellent green salad mixed with quartered tomatoes, crumbled bacon, and French dressing. A chilled white wine is also good.*

CHICKEN AND TOMATO SOUP
WITH WINE

Serves: 6 (or more)
Time: About 35 minutes

3 cups cooked chicken
8 ounces tomato juice
3 cloves garlic
3 teaspoons sugar
salt and freshly ground pepper
  to taste

½ teaspoon basil
4½ cups chicken broth
6 ounces dry white wine
sour cream (optional)

Dice the cooked chicken. Combine tomato juice, garlic, sugar, salt and pepper, and basil and boil briskly for 15 minutes. Strain and add the diced chicken, chicken broth and wine and boil for 20 minutes or more. Serve hot. (With a dab of sour cream on top, this makes an excellent cold soup for summer.)

/ Serve with chef's salad and a dry white wine.

CHICKEN BALLS WITH PUMPKIN PUREE

Serves: 4 to 6
Time: About 45 minutes

1½ pounds boned fresh
  chicken
2 eggs (beaten)
3 tablespoons butter
2 medium onions (chopped)
2 small stalks celery (diced)
1½ tablespoons flour

1 teaspoon salt
¼ teaspoon ground ginger
½ teaspoon nutmeg
2 cups pumpkin purée
3½ cups chicken broth
1 cup light sweet cream

Have butcher grind the chicken. Place in a bowl and mix in the beaten eggs. Cool in refrigerator 15 to 20 minutes. Meanwhile, melt the butter in a large kettle and add the onions and celery. When limp but not brown, push to one side and blend in flour. Now add salt, ginger, nutmeg, and pumpkin. Stir well and gradually add the chicken broth. Bring to a boil, stirring constantly. Reduce heat and simmer while you make chicken balls. These should be about the size of a walnut—form them with dampened hands. Add these to the simmering pumpkin and cook about 20 minutes. Remove chicken balls and strain the sauce. Return sauce to kettle and gradually add the cream, then return the balls. Simmer for about 15 minutes. As a novelty, serve in a scooped out pumpkin (heated in oven).

⁄ Serve with broccoli, escarole salad, and a chilled Rhine wine.

PAPRIKA CHICKEN BALLS WITH SHERRY

Serves: 4 (or more)
Time: About 40 minutes

4 chicken breasts (boned)
1 large onion (minced)
1 clove garlic (crushed)
2 eggs (beaten)
salt and freshly ground pepper
8 saltine crackers (crumbled)
½ cup milk
2 tablespoons oil
1 large onion (sliced)
½ cup sherry wine
paprika
2 tablespoons parsley (chopped)

Ask butcher to grind the chicken breasts. Place in large bowl and add the onion, garlic, eggs, and salt and pepper. Meanwhile, soak crumbled crackers in milk. Drain and add to chicken mixture and beat all ingredients until fluffy. Place in refrigerator for 15 to 30 minutes. Dampen hands with oil or water and form chicken into walnut-size balls. Heat the oil in a large skillet and brown onion slices lightly on both sides. Place the meatballs over onion and add sherry. Sprinkle well with paprika, reduce heat, cover skillet, and simmer for about 30 minutes. Should bottom of pan dry out, add a bit more sherry. Garnish with chopped parsley.

⁄ Serve with broiled tomatoes with sesame seeds, watercress salad, and a chilled Moselle wine.

EASY CHICKEN CUTLETS

Raoul Lionel Felder is a noted attorney and author. He is also a customer of ours whose wife devised this interesting recipe for chicken cutlets.

Serves: 4
Time: About 16 minutes

dark meat from 2 chickens (boned)
4 tablespoons butter
1 large onion (chopped)
¼ teaspoon tarragon
½ teaspoon salt
2 tablespoons parsley (chopped)
2 cups herb bread (Pepperidge Farm)

Have chicken meat ground. Place butter in frying pan and sauté the onions until transparent. Pour this mixture into a bowl and add the chopped chicken, tarragon, salt, and parsley. Preheat oven (400° F.). Meanwhile, place the bread in a blender. When well crumbled, add to chicken mixture and mix well with hands. Then form into plump patties and place on flat pan under broiler. It is important that the heat is only 400°— not the full broiler heat. Cook for 8 minutes on each side.

⌐ Serve with fluffy rice, green salad (tossed at table), and Pouilly-Fuissé white Burgundy.

CHICKEN CROQUETTES

Serves: 4
Time: About 30 minutes

2 cups cooked chicken (ground)
2 tablespoons butter
1 medium onion (minced)
1 egg (beaten)
½ cup sour cream
salt and freshly ground pepper
1 tablespoon parsley (chopped)
½ cup corn flakes (crumbled)
¼ pound butter
8 lemon wedges

Place chicken in large bowl and bring to room temperature. Melt butter in small frying pan and sauté onion until golden. Remove from heat and cool. Mix beaten egg with sour cream and salt and pepper. Add to chicken and then pour in onion with butter it was cooked in. Add the parsley and mix well. Place bowl of chicken mixture in refrigerator for 15 to 30 minutes. Dampen hands and form chicken into 2-inch pyramids and roll these in crumbled corn flakes. Heat butter in large skillet and, when bubbling, add the croquettes. Brown them well on all sides. Drain them on paper towels and serve with lemon wedges.

ꜰ *Serve with buttered peas, creamed onions, a rosé wine.*

CHICKEN SALAD NO. 1

Chicken salad is just about as American as hot dogs and apple pie. It used to be a favorite with women's bridge clubs. Now it has become a favorite with families—children (especially teens) seem to love it. It is easy to make and is a wonderful way to use chicken leftovers. Chicken salad can have many variations—it does not have to be the usual chicken, celery, and mayonnaise. It can be varied in many ways, so let your imagination go to work. Here are three variations we like very much.

Serves: 4

1 cup leftover chicken (diced)	2 tomatoes (cut in eighths)
2 stalks celery (stringed and diced)	salt and freshly ground pepper
	4 tablespoons mayonnaise
2 carrots (scraped and diced)	crisp romaine or other lettuce
2 apples (peeled, cored, diced)	paprika

Place diced chicken in large bowl. Add the celery, carrots, and apples. Be sure you drain the tomato pieces on paper towels before adding them. Sprinkle this mixture with salt and pepper, then add the mayonnaise and mix everything well. If the mixture seems a bit too thick, add a little lemon juice. Serve on crisp green leaves and sprinkle with paprika.

ꜰ *This is a hearty dish, so have only a cup of consommé first.*

CHICKEN SALAD NO. 2

Serves: 6

2 cups cooked chicken (diced)
2 tablespoons Parmesan cheese
  (grated)
½ cup olive oil
3 tablespoons tarragon vinegar
½ teaspoon dry mustard

4 dashes Worcestershire sauce
salt and freshly ground pepper
crisp salad greens (broken up
  or chopped)
1 cup croutons

Place chicken in a large bowl and mix with the cheese. Meanwhile, mix oil, vinegar, mustard, Worcestershire, and salt and pepper. Stir this until it is well blended. Pour over salad. Stir all of this well and refrigerate until ready to use. At the last minute, add greens. Stir thoroughly and serve on chilled plates. Garnish with croutons.

/ While the salad is chilling, you could have cups of hot bouillon. And with the salad, serve crisp, toasted garlic bread—and wine, of course, dry, white, and chilled.

CHICKEN SALAD NO. 3

This is a salad that an artistic cook should love. It is so full of colorful ingredients that can be arranged attractively in patterns in a large glass bowl—which will look like a flower combination on the party table before it is mixed with the special dressing.

Serves: 6 to 8

3 cups cooked chicken
1 package frozen string beans
  (cooked and chilled)
4 stalks celery (stringed and
  chopped)
2 green peppers (cut in rings)
1 pint fresh cherry tomatoes
1 small can flat anchovies
10 black olives (pitted and
  sliced)
10 stuffed green olives
5 new potatoes (cooked and
  sliced)

2 medium Bermuda onions
  (sliced)
¼ cup parsley (chopped)
¼ cup scallions (chopped)
6 hard-cooked eggs (quartered)
crisp romaine leaves
¾ cup salad oil
2 teaspoons dry mustard
salt and freshly ground pepper
2 cloves garlic (minced)
2 tablespoons wine vinegar

Cut the chicken in bite-size cubes and arrange with the vegetables and eggs in patterns in a large glass bowl. Then stick the romaine (stem end down) around the edge of the mixture (this will resemble an Indian feathered hat). Put in refrigerator to chill. Meanwhile, place oil, mustard, salt and pepper, garlic, and vinegar in blender and mix well. Serve this in small bowl.

┐ *Serve with jellied consommé, crisp French bread, and a dry white wine.*

CREAMED CHICKEN AND SWEETBREADS

This is a great party dish, rich and delicious. The cooking time is short and, inasmuch as it is cooked in a double boiler, it can be held if the guests are not ready to eat. The preparation is what takes time. The chicken must be cooked and the sweetbreads must be parboiled and deveined (see pp. 242–243 for sweetbread preparation).

Serves: 8 to 10
Time: About 25 minutes

1 5-pound chicken (cooked)	2 cups light sweet cream
2 sets veal sweetbreads (parboiled)	2 egg yolks (beaten)
	salt and white pepper
2 tablespoons butter	¼ cup dry sherry
2 tablespoons flour	3 tablespoons parsley (chopped)

Cut chicken into bite-size cubes and dice the sweetbreads. Melt the butter in top of double boiler. Blend in flour, stirring constantly, and gradually add 1¾ cups of cream. Stir until slightly thickened. Mix ¼ cup of cream with the egg yolks and slowly add to the sauce, stirring constantly. Then add the chicken and sweetbreads and season to taste with salt and white pepper. Cover the pan and allow to steam until ready to serve. Add sherry just before serving and garnish with parsley.

┐ *Fill patty shells with creamed chicken and serve buttered peas on the side—and because it's a party, champagne in chilled glasses.*

ROASTED CAPON NO. 1

Serves: 6
Time: 2½ hours

6- to 7-pound capon
stuffing
3 tablespoons butter
1 small onion (finely grated)
2 cloves garlic (grated)

3 teaspoons mushrooms (finely chopped)
salt and freshly ground pepper
½ cup white wine

Wash capon inside and out and pat dry with paper towels. Capon may be roasted without stuffing, but a good stuffing certainly takes the place of a starchy vegetable (see pp. 271 to 278 for varieties of stuffings). Stuff the bird and truss it. Preheat oven (325° F.). Meanwhile, melt butter over low heat and add onions, garlic, mushrooms, and salt and pepper. Heat through but do not brown the ingredients. Add the wine and stir until all is well warmed. Brush some of this mixture all over capon. Now place bird (breast side down) on rack in open roasting pan and bake for 1 hour, basting with some of the sauce every 15 minutes. Turn bird breast side up and continue roasting for 1½ hours. Baste frequently during this period. Turn off oven heat and allow capon to rest in oven undisturbed for 10 minutes.

↑ *Tip: If capon is baked without stuffing, reduce cooking time by 15 minutes.*
↑ *If you have not used a stuffing, serve creamy mashed potatoes. Otherwise, just include your favorite green vegetable—and your favorite chilled white wine.*

ROASTED CAPON NO. 2

Serves: 6
Time: 2¼ hours

6- to 7-pound capon
stuffing
3 tablespoons butter
2 tablespoons flour

2 tablespoons juice from
seedless grapes
salt and pepper

Wash capon inside and out and dry well. Stuff with your favorite stuffing (see pp. 271 to 278) and truss. Preheat oven (375° F.). Meanwhile, melt butter in pan and gradually add flour, stirring constantly. Then add grape juice and salt and pepper. When this has become a pastelike mixture, spread it over the bird. Place capon on rack in uncovered pan in oven. Bake for half an hour, then cover loosely with aluminum foil and reduce heat to 325° F. Continue roasting for about 1¾ hours. Remove foil during last 15 minutes of cooking and baste with pan drippings. Turn off oven heat and leave capon in oven for about 10 minutes.

ɤ *Tip: If capon is roasted without stuffing, reduce cooking time by 15 minutes.*
ɤ *Serve with creamed spinach and a crisp endive salad—also a rosé wine*
ɤ *Variation: A 5-pound roasting chicken can be cooked in the same way. But after 2 hours of cooking, test for doneness with a fork. It will probably take 15 or so minutes less cooking time.*

BONELESS CAPON WITH CHESTNUT STUFFING

This is a great delicacy that can be served hot for dinner or cold for a buffet. When stuffed properly it looks like a well-rounded bird, and the host creates a stir when he slices it straight through so that each guest receives a large oval (½ inch or thicker) of capon with a center of stuffing. Boning a capon is one of our specialties, and we recommend Chestnut Stuffing for Poultry, but many other stuffings are excellent for this party dish (see pp. 271 to 278).

Serves: 6 (generously)
Time: 2¼ hours

8- to 9-pound capon	Chestnut Stuffing for Poultry
salt and freshly ground pepper	(pp. 274–275)
	1 tablespoon chicken fat (soft)

Have butcher bone capon without cutting skin in back. Be sure he takes out all bones except wing bones. Preheat oven (350° F.). Wash capon and dry carefully. Sprinkle with salt and pepper inside and out. Now stuff the bird so that it has a rounded shape and sew up the cavity. Smooth the chicken fat over the whole bird. Place on a rack in uncovered roasting pan and bake for 2¼ hours. Baste with pan drippings at least 4 times during roasting. Remove from oven and let stand for 20 minutes. Remove sewing string and serve.

ｧ *Serve with buttered peas and pearl onions, hearts of palm salad, and champagne.*

ROAST TURKEY HEN NO. 1

There are various ways to roast a turkey. Here are two that we particularly like. Of course, all turkeys should contain a luscious stuffing (you will find several varieties on pp. 271 to 278). Roast Turkey Hen No. 1 requires a session of marinating, and No. 2 has a top-of-the-stove precooking period. We'll leave it up to you to decide which is best.

Serves: 8 to 10
Time: About 3½ hours

12- to 14-pound hen turkey	3 garlic cloves (crushed)
¾ cup oil	¼ teaspoon paprika
¾ cup wine vinegar	1 tablespoon onion juice
salt and freshly ground pepper	turkey stuffing (see pp. 271
3 teaspoons Worcestershire	to 278)
sauce	

Wash turkey well inside and out, then dry. Now blend the next 7 ingredients. Pour this marinade over turkey in a large pan and let stand for 8 hours (this can be at room temperature or in refrigerator, but the marinade should be basted over turkey during this waiting period at least 4 times). Before roasting, be sure turkey is at room temperature. Preheat oven to 325° F. Drain the turkey well, reserving the marinade. Fill cavity with your favorite stuffing, close cavity with skewers and string, and tie legs together. Place on rack in open roasting pan. Roast for 2 hours and brush often with the marinade you have saved. Now raise the heat to 350° F. and continue roasting for another 1½ hours. During this last cooking period, baste only one time

—using every speck of marinade you have left. Before removing from oven, test the bird with a sharp 2-tined fork to test doneness. If fork does not come out easily, cook a few more minutes, but do not increase the heat.

⸙ *Pick a vegetable that contrasts with the stuffing you have used. For example, if the stuffing is starchy, do not serve potatoes. Instead, have your favorite green vegetable. Cranberry sauce or jelly is almost a must. Creamed onions are another favorite, and so is pureed pumpkin. Any wine would be delightful, but for a particularly festive meal, bring out the champagne.*

ROAST TURKEY HEN NO. 2

Serves: 8 to 10
Time: About 3½ hours

12- to 14-pound turkey hen	1 cup water
salt and freshly ground pepper	2 cups apricot juice
stuffing (pp. 271 to 278)	flour (for gravy)

Wash turkey inside and out and dry well. Then sprinkle with salt and pepper. Pick your favorite stuffing and fill cavity of bird. Then tie up cavity with skewers and string and tie legs together. Place turkey on a rack in large baking pan. Add water and apricot juice to bottom of pan and place over low heat on top of stove. Cover and simmer for about an hour. Take a look from time to time and baste with juices on bottom of pan (add more water if it has dried out). Preheat oven to 350° F. and place turkey in same pan (but uncovered) in oven. Bake for 2½ hours, basting about every half-hour. Of course, add more water if bottom of pan dries out. Remove bird to hot platter. Also, remove rack and put pan on top of stove over medium heat. Add flour a little bit at a time and stir constantly so that gravy will not lump. Add more water if necessary.

⸙ *The stuffing is your clue about what vegetable to serve. If it will fit in with your plans, hominy with lots of butter is an old southern favorite, and so are pureed chestnuts. Add a festive wine.*

SCALLOPINI OF TURKEY BREAST MARSALA

Serves: 6
Time: 10 to 12 minutes

Almost everyone thinks of "scallopini" as meaning veal. However, many types of meats (or birds) can fit into the scallopini category if they are thinly sliced or pounded. Here is an unusual one that is delicious.

3-pound turkey breast (from 10-
 to 12-pound turkey)
2 tablespoons flour
salt and freshly ground pepper
3 tablespoons butter

¼ cup chicken broth
½ cup Marsala wine
1 tablespoon butter
1 can button mushrooms

Have butcher bone turkey breast and cut into very thin slices. Pound these between wax paper until they are even thinner. Then dip into flour that has been seasoned with salt and pepper. Heat the butter in a large skillet. When it bubbles, add the turkey slices and sauté for 6 minutes on each side. As they are browned, place slices on a hot platter. To the skillet, add chicken broth and wine. Heat thoroughly and then add the tablespoon of butter and the drained mushrooms. Mix well. Actually, the flour from the turkey pieces should have left enough residue to make the added liquids thicken somewhat. However, if the juice is not of a gravylike consistency, cool about 2 tablespoons of the broth and stir in a teaspoon or so of arrowroot. When the sauce is the consistency you wish, pour it over the turkey and serve.

⁊ *Serve with tiny noodles, asparagus tips, and a medium-dry red wine.*

TURKEY SCALLOPINI ROLLS (WITH PROSCIUTTO AND SWISS CHEESE)

Serves: 4 to 6
Time: About 25 minutes

8 slices turkey scallopini
8 slices prosciutto ham
8 slices Swiss cheese
  (thinly cut)
¾ cup seasoned breadcrumbs
¾ cup flour

1 egg
3 tablespoons water
¼ cup oil
1 tablespoon flour
1 cup bouillon
watercress—for garnish

As in the previous recipe, have butcher cut the breast of turkey into very thin slices and pound them until they are as thin as veal scallopinis. Place these thin slices on waxed paper. On top of each, arrange a slice of prosciutto and then a slice of cheese. Roll these up and secure them with toothpicks. Preheat oven (350° F.). Combine the seasoned breadcrumbs and flour. In another dish, beat together the egg and water. Now dip the rolls into the breadcrumb mixture, then into the egg solution, then once again into the breadcrumbs. Be sure that they are well covered. Heat the oil in a large heavy skillet. When sizzling, add the rolls and brown on all sides until golden. Remove the rolls and place in a shallow baking dish. Leave the skillet over the heat and gradually add and stir the tablespoon of flour into the drippings. When this is well mixed, slowly add the bouillon, stirring all the time. When this mixture is bubbling and has thickened slightly, spoon it over the rolls. Place in oven for 15 minutes or until done. Before serving, be sure to remove the wooden picks and then garnish with watercress.

＊ *Serve with green noodles and broiled tomatoes. A chilled Chablis will add quite a bit to this menu.*

ROAST TURKEY BREAST

Serves: 2 or 3
Time: 1½ hours

breast from 10- to 12-pound
  turkey
½ lemon
Bread Stuffing for Poultry
  (p. 272)

4 tablespoons butter (soft)
coarse salt and pepper
1 cup chicken broth

You can have the turkey breast boned when you buy it, but you can easily do it yourself. Just use a very sharp small knife and hold it close to the breast bone and move it gently so that the meat stays firm. Preheat oven (350° F.). Now rub meat and skin with lemon. Place stuffing in center of breast and wrap breast around it. Insert skewers and wind string around them to make a crisscross closure. Place turkey breast, skewer side down, on rack of small roaster after brushing a bit of the butter over rungs of rack to keep turkey from sticking. Spread rest of butter over top of turkey and sprinkle generously with coarse salt and freshly ground pepper. Roast for about 20 minutes—by this time the butter will have drizzled down to pan. Add chicken broth and continue cooking for an hour or so, basting from time to time.

ꝉ *Serve with a crisp green vegetable or salad and a nice Sauterne.*

TURKEY STEW WITH SHERRY

There are three recipes in this book that use turkey breasts. Here is one that will delightfully take care of the rest of the bird.

Serves: 8
Time: About 2¼ hours

10- to 12-pound turkey (breast
  removed)
1 cup flour
salt and freshly ground pepper
¼ pound butter
2 large onions (chopped)

4 carrots (chopped)
1 bay leaf
3 tablespoons parsley (chopped)
¼ cup water
¼ cup sherry

Ask butcher to cut turkey into 2-inch squares. He can do this readily with his bandsaw so that meat and bone are intact within the 2-inch pieces. Wash turkey pieces and dry well on paper towels. Dip each piece into flour flavored with salt and pepper. Melt butter in a large skillet and brown pieces on every side over high heat. Remove turkey pieces to paper towels and add onions, carrots, and bay leaf to remaining butter (add more butter if necessary). When they have softened a bit, stir in the parsley. Now return the turkey pieces and add the water and sherry. When juice has bubbled, reduce heat, cover pan, and simmer for about 1½ hours. Lift the cover during this cooking period and see if the juice has dried out. In that case, just add a bit more sherry. Then remove lid and simmer for about another half-hour. When the meat falls off a fork easily, you will know that it is well done. Remove bay leaf. Before serving, adjust the seasoning and either thicken (with arrowroot and water) or dilute (with a bit more wine).

⸹ *Serve with fluffy mashed potatoes, buttered string beans, and a cool light red wine.*

TURKEY HASH

Serves: 4
Time: About 35 minutes

2 cups cooked turkey (chopped)	½ cup green peppers (chopped)
1 cup leftover stuffing	½ cup onions (chopped)
salt and freshly ground pepper	½ cup leftover turkey gravy
3 tablespoons butter	

Mix turkey and stuffing in bowl and season to taste with salt and pepper. Melt butter in frying pan and sauté peppers and onions. When slightly brown, add turkey mixture and gravy and stir well. Cook over low heat for 25 minutes. Increase heat last 10 minutes so that hash will brown and become nicely crusted. Shake pan occasionally to prevent sticking.

⸹ *Serve with buttered lima beans and creamed onions.*

TURKEY SOUP

Serves: 6
Time: 1¼ hours

carcass from roast turkey
wings from roast turkey
turkey giblets (cut up)
3 carrots (cut in halves)

1 large onion (cut in half)
2 stalks celery (quartered)
salt and freshly ground pepper
2 teaspoons chicken-soup mix

Place turkey carcass, wings, and giblets in large pot and cover with 2½ quarts water. Bring to boil and add carrots, onion, and celery. Cover pot and reduce heat. Simmer for half-an-hour, then add salt and pepper to taste and chicken-soup mix. Continue simmering for 45 minutes. Remove carcass and discard. Cut meat from bones in small pieces. Return meat to soup and discard bones.

✝ *Serve with chef's salad and crunchy French bread.*

ROAST DUCK FLAMBÉ NO. 1

Serves: 4
Time: About 2½ hours

4- to 5-pound duckling
1 lemon (quartered)
salt and pepper
16 prunes

16 apricots
1 orange (sliced in 4 places)
½ cup Cointreau

Wash duck well and dry. Rub inside and out with lemon and place rinds in cavity. Sprinkle with salt and pepper and prick skin all over with sharp-tined fork to allow the fat to escape. Preheat oven (450° F.). Meanwhile, cover prunes and apricots with cold water, bring to boil, and cook for 5 minutes. Drain well and place in cavity. Tie legs together. Place on rack of roasting pan and cook uncovered for 25 minutes. Reduce heat to 350° F. Remove duck from oven, lift out rack with duck, and place on brown paper. Discard all fat and wipe pan clean with paper towels. Return duck with rack to pan and cook for 1 hour. Place orange slices on top of duck and cook for about 1 hour. Just before serving, heat Cointreau, pour over duck, ignite, and bring to table flaming.

🍴 *Serve with mashed sweet potatoes, mixed green salad, and a dry Bordeaux wine.*

ROAST DUCK FLAMBE NO. 2

Serves: 4
Time: About 2½ hours

4- to 5-pound duck
salt and freshly ground pepper
4 tablespoons butter
8 small white onions
4 carrots (sliced lengthwise)

1 orange (skinned and sliced)
½ cup currant jelly
½ cup dry red wine
½ cup brandy (heated)

Preheat oven (350° F.). Wash duck inside and out and dry thoroughly. Sprinkle inside and out with salt and pepper. Prick skin all over. Place on rack in roasting pan and bake for 2½ hours, draining off fat from time to time. When duck has cooked for 2 hours, melt butter in skillet and sauté onions and carrots over medium heat. When carrots slip off from fork easily, add orange slices, jelly, and red wine. Allow sauce to bubble until it has thickened slightly. Meanwhile, place duck on a hot platter. Surround it with carrots, onions, and orange slices in an attractive pattern. If juice has thickened too much, add a bit more wine, then pour over duck and vegetables. Now pour the heated brandy over the duck and flame.

🍴 *Serve with buttered broccoli, spicy apple sauce, and a chilled Burgundy wine.*

ROAST DUCK WITH ORANGE

Serves: 4
Time: 2 hours

5-pound duck
1 cup fresh orange juice
3 tablespoons almonds
  (chopped)

2 teaspoons flour
salt and freshly ground pepper
3 oranges (quartered)

Puncture skin of duck in various places with a sharp 2-tined fork to release fat. Now place duck in a large pan with enough boiling water to cover. Boil briskly for half an hour. Preheat oven (350° F.). Meanwhile, combine orange juice, almonds, flour, and salt and pepper. Remove duck from boiling water and drain. When cool enough to handle, rub orange mixture inside and outside of duck. Place on a rack in a roasting pan and surround duck with orange sections. Roast uncovered for 1 hour, basting each 30 minutes. Raise oven heat to 375° F. and continue roasting for another half-hour, basting frequently.

↙ Serve with fried rice (with garlic butter), Brussels sprouts, and a Burgundy wine.

ROAST DUCK WITH APPLES

Serves: 4
Time: About 2½ hours

4- to 5-pound duck
2 tablespoons brandy
coarse salt
1 teaspoon thyme

1 large onion (quartered)
2 large apples (peeled,
  quartered)
5 strips bacon

Wash duck inside and out and dry thoroughly. Prick the skin all over. Here is a trick: Flambé the duck before it has been cooked—it gives an interesting and subtle flavor. Heat the brandy in a soup ladle, light with a match, and pour over duck. When flame has burned out, sprinkle duck with salt and thyme. Fill cavity with onion and apples. Then close cavity with skewers and string. Preheat oven (350° F.). Place the bacon slices over breast of duck and place on rack in uncovered baking pan. Roast for 30 minutes, remove bacon, then continue roasting for 2 hours. Be sure to drain off fat from time to time.

✝ *Serve with baked sweet potatoes, buttered broccoli, and a medium-dry red wine.*

ROAST DUCK WITH FRUIT AND LIQUEUR SAUCE

Edward H. Benenson, a noted New York realtor, is president of Chaine des Rotisseurs and chief of the New York and international chapter of Confrérie des Chevaliers du Tastevin. He is also a great cook. Here is one of his favorite recipes: roast duck that "borrows a bit of the Chinese method with a great French sauce, touched with something of my own."

Serves: 2 to 4
Time: 2 hours

5- to 6-pound tender duck
garlic salt
1 "toe" garlic (slivered)
1 pint apple cider
2 tablespoons honey
3 halves canned peaches
 (sliced)

¼ pound butter
1 cup heavy peach syrup
small jar bar-le-duc (red
 currant jam)
1 cup dry white wine
½ cup Grand Marnier
1 pony cognac

Preheat oven (400° F.). Clean the duck well but do not dry it. Rub garlic salt in the cavity. Place paper-thin slivers of garlic under the outside skin in about six different places. Put duck on rack in open roasting pan and place in preheated oven. Meanwhile, combine the cider and honey and baste duck with it—this helps to crisp the skin and make it beautifully brown. After half an hour reduce heat to 350° F. and roast for another 1½ hours. It is very important to drain off duck fat from bottom of pan while it is baking. This grease prevents proper circulation of oven heat and tends to make the bird greasy. During cooking period, keep basting with cider and honey mixture—using a brush—and watch the way the skin mellows in color. Also, make sure to expose the bottom of the duck during roasting for about 10 minutes and baste so that bird will brown evenly all over. While duck is cooking, make the sauce. Add the sliced peaches to heated butter in a large saucepan. Then add peach syrup and currant jam and bring to a slow froth, stirring constantly. Now add the white wine, Grand Marnier, and cognac. Stir until the mixture is slightly thickened and serve hot with the duck. If the duck is not quite finished, keep sauce hot in a double boiler.

⨍ *Serve with wild rice, a crisp green vegetable (fiddle ferns if available), and a good claret (at room temperature).*

BRAISED CHINESE DUCK

Serves: 4
Time: About 2¾ hours

5-pound duck	3 tablespoons sherry
½ cup Chinese sauce (Hoisin)	5 tablespoons oil
4 scallions (chopped)	2 cups water
2 slices fresh ginger root	1 tablespoon sugar

Wash duck inside and out and dry thoroughly. Meanwhile, combine Chinese sauce, scallions, ginger root (if you can't get the fresh root, use ½ teaspoon powdered ginger), and sherry. Now rub part of the Chinese-sauce mixture inside and outside of duck (reserving the rest) and let duck stand at room temperature for 1 hour. Heat oil in large skillet or Dutch oven and brown duck lightly on all sides. Drain off all fat from pan and add remaining sauce mixture and water. Bring to a boil, reduce heat, and simmer, covered, for 1¼ hours. Add sugar and resume simmering for another half-hour covered. Should sauce dry out, add a little more sherry.

⨍ *Serve with fluffy rice, buttered Chinese cabbage, and a medium-dry red wine.*

BRAISED DUCK WITH WINE

Serves: 4
Time: About 2 hours

5-pound duck (cut in eighths)
1 cup Sauterne
¼ cup Marsala
1 orange (sliced)
½ lemon (sliced)
2 tablespoons seedless grapes
   (crushed)
½ cup water
¾ teaspoon celery seed
½ teaspoon marjoram
½ teaspoon thyme
salt and freshly ground pepper
1 cup flour
5 tablespoons oil
12 small white onions (sliced)
12 mushroom caps
1 tablespoon cornstarch

Wash duck pieces and dry well. Combine next 10 ingredients (up to flour). Add duck to this marinade and refrigerate for 24 hours. Remove duck and dry well on paper towels and strain the marinade, reserving sauce. Dip duck pieces in flour and put in large skillet or Dutch oven sizzling with oil. Brown well on all sides. Then add onions and mushrooms and cook slightly. Drain off all fat and add strained marinade. Bring to a boil, reduce heat, and simmer covered for 1½ hours, stirring every 30 minutes. Remove cover and simmer 15 minutes more. Meanwhile, preheat broiler. Remove duck pieces to shallow baking pan and place under broiler for 10 minutes, fat side up. This should make the skin become crisp. Meanwhile, cool a bit of the sauce and mix with cornstarch. Add to sauce and stir until thickened. Pour sauce over duck after 5 minutes of broiling.

⨍ *Serve with mashed sweet potatoes, buttered broccoli, and a medium-dry red wine.*

BRAISED DUCK WITH GIBLET SAUCE

This recipe is in two parts—the actual cooking of the duck and the concocting of the giblet sauce. Therefore we will list the ingredients in two sections. (Giblets consist of the gizzard, heart, and liver.)

Serves: 8
Time: About 2 hours

The Duck

2 5-pound ducks (skinned and quartered)
½ cup flour
salt and freshly ground pepper
½ cup oil

6-ounce can frozen orange-juice concentrate (thawed)
⅓ cup currant jelly
2 oranges (sliced)
2 tablespoons parsley (chopped)

The Giblet Sauce

gizzards and necks from the 2 ducks
½ cup onions (chopped)
4 scallions (chopped)

¾ teaspoon salt
hearts and livers from the 2 ducks
2 teaspoons flour

Be sure that all fat has been removed with the skin from the ducks. Coat duck quarters with flour mixed with salt and pepper. Sizzle oil in large heavy skillet and brown on all sides. It is best to do only a few pieces at a time. As each is browned, drain on paper towels.

Meanwhile, place gizzards and necks in boiling water to cover, add onions, scallions, and salt, and simmer for 30 minutes. Add hearts and livers and simmer for about 15 minutes. Drain, reserving cooking liquid. Chop giblets and "dismantle" necks and chop meat. Now mix flour into the fat left over from browning the duck. Gradually add the giblet stock and then the giblets and the chopped neck meat—this is the sauce you will need soon.

Place browned duck quarters in Dutch oven over medium heat. Add 1 cup of giblet sauce. When bubbling, gradually add the orange-juice concentrate, currant jelly, and remaining giblet sauce. Cover pan and simmer for about 1½ hours. Be sure to watch it closely, however, so that you can adjust the thickness of the sauce. If it is too watery, you can always cool a bit of the juice and add some arrowroot or cornstarch to it. And if it is too thick, you can add a little water, wine, or orange juice. Now, when you feel the dish is perfection, serve topped with orange slices with a sprinkle of chopped parsley.

⅟ *Serve with long-grain rice, a tomato and lettuce salad, and your best dry red wine.*

DUCK CASSEROLE

Serves: 4
Time: About 2 hours

4- to 5-pound duck (cut in pieces)	3 tablespoons butter
1 lemon (quartered)	thyme
flour	1 medium onion (chopped)
salt and pepper	1 cup dry red wine
	1 cup consommé

Wash duck pieces and dry well. Rub with lemon and let stand as long as 15 to 30 minutes. Shake each piece in bag with flour and salt and pepper. Preheat oven (350° F.). Dust off any loose flour, and place pieces in skillet with sizzling butter. Brown well on all sides. Place on paper towels to drain off all fat, then put into casserole. Add good pinch of thyme, chopped onions, wine, and consommé. Cover and bake for 1¾ hours. Inspect toward end of cooking time. If sauce has dried out, add more wine.

⅟ *Serve with green peas and onions, tossed salad, and a dry red wine.*

STEWED DUCK WITH CHESTNUTS

Serves: 4 to 6
Time: About 1½ hours

4- to 5-pound duck
1 bay leaf
4 tablespoons butter
1 cup mushrooms (sliced)

1 cup chestnuts (chopped)
salt and freshly ground pepper
4 tablespoons soy sauce
1 cup sherry wine

Have duck cut into eighths. Place parts in boiling water and add bay leaf. Cover and simmer for about 1 hour. Discard bay leaf and remove duck pieces and drain, then pat them dry on paper towels. Now melt butter in large skillet. When bubbling, brown duck on all sides. Push duck to one side, reduce heat, and add mushrooms. Stir them until wilted and add chestnuts, salt and pepper, and soy sauce (use only a small amount of salt, as soy sauce is a strong, salty seasoning). Stir everything together and gradually add sherry wine. Cover and simmer for about 30 minutes.

⸰ Serve with fluffy rice, endive salad, and a medium-dry red wine.

DUCK SALAD WITH FRUIT AND NUTS

Serves: 6

2 cups cooked duck (diced)
2 cups mandarin orange
   sections (drained)
1 cup grapefruit sections
   (drained)

¾ cup mixed nuts (chopped)
½ cup French dressing
crisp romaine or other lettuce

Mix first 5 ingredients and chill thoroughly. Serve on crisp green leaves.

⸰ This makes a delightful luncheon dish. First you could have a hot bouillon, then melba toast and a very dry white wine with the salad.

ROCK CORNISH GAME HEN—ROASTED

Serves: 4
Time: About 55 minutes

2 Rock Cornish game hens	Savory Kasha Stuffing for Poultry
½ lemon	(p. 274)
salt and pepper	4 strips bacon
	¾ cup red wine

Rub birds inside and out with lemon and sprinkle well with salt and freshly ground pepper. Preheat oven (450° F.). Fill cavities with stuffing. Close opening with skewers and lace with string crisscrossed around them, also tying legs. Place birds, breast side up, on rack in open roasting pan and cover breasts with bacon. Cook for 15 minutes. Reduce heat to 325° F. and add red wine. Roast for 35 to 40 minutes, basting occasionally and adding more wine if needed.

⸫ *Serve with asparagus in lemon butter and a Burgundy wine.*

ROCK CORNISH GAME HEN WITH HERB AND SHALLOT BUTTER

Serves: 4
Time: About 1 hour

2 Rock Cornish game hens	1 teaspoon rosemary
salt and freshly ground pepper	1 teaspoon thyme
¼ pound butter	2 tablespoons chopped parsley
5 shallots (minced)	

Preheat oven (350° F.). Cut each hen in half, sprinkle all over with salt and pepper, and bring to room temperature. Melt butter in small fry pan and add shallots, rosemary, and thyme. Stir well and let bubble for about 5 minutes, then remove from heat. Place hens in flat baking pan, skin side up, and baste with the butter sauce. Bake for 50 minutes, basting every 15 minutes. Garnish with parsley.

⸫ *Serve with wild rice, leaf spinach, and a rosé wine.*

ROCK CORNISH GAME HEN BREASTS (KIEV)

Serves: 4
Time: 40 minutes

4 breasts of Rock Cornish game hens (boned)	1 egg (beaten)
salt and freshly ground pepper	4 tablespoons seasoned breadcrumbs
4 ounces butter	

Wash and then dry hen breasts thoroughly, then sprinkle with salt and pepper. Preheat oven (350° F.). Place an ounce of butter in each breast and close with a skewer. Dip each piece in egg and roll in breadcrumbs. Place pieces in slightly greased, shallow baking pan and bake for 40 minutes. Turn breasts after 20 minutes.

✓ Tip: Save legs, wings, and backs of these birds and use in Braised Rock Cornish Game Hen (this page).
✓ Serve with braised English celery, yams with shredded coconut, and a medium-dry white wine.

BRAISED ROCK CORNISH GAME HEN

Serves: 4 to 6
Time: About 50 minutes

4 Rock Cornish game hens (with breasts removed)	6 scallions (chopped)
flour	½ pound mushrooms (sliced)
salt and freshly ground pepper	1 teaspoon oregano
¼ pound butter	½ cup dry white wine
	2 tablespoons parsley (chopped)

Cut legs and wings away from back and cut back in half. Wash pieces and dry thoroughly. Dip each piece into flour seasoned with salt and pepper. Heat butter in large skillet. When sizzling, add hen pieces and brown on all sides. Remove to paper towels to drain. Now add scallions, mushrooms, and oregano to remaining butter (add more if necessary). When these have wilted, return the hen pieces and add the wine. Reduce heat, cover pan, and simmer for about 35 minutes. Lift lid from time to time and add more wine if juice has dried out. Just before serving, remove any meat that clings to back and discard the bones. Garnish with parsley.

⁊ *Tip: This recipe uses the parts of the bird that are not used in Rock Cornish Game Hen Breasts (Kiev) (p. 228).*
⁊ *Serve with fluffy rice, tiny Belgian carrots, and a chilled Sauterne wine.*

ROASTED GUINEA HEN WITH WILD RICE STUFFING

Serves: 4
Tim: 1½ hours

2 guinea hens
salt and freshly ground pepper

Savory Wild Rice Stuffing for Poultry (p. 273) or Wild Rice Stuffing with Grapes for Poultry (pp. 272–273)

Wash hens and then dry thoroughly with paper towels. Preheat oven (350° F.). Sprinkle the birds with salt and pepper. Fill cavities with stuffing and truss them (skewers laced with string). Place each bird in the center of a large sheet of aluminum foil and wrap them tightly, folding over each end. Place them, breast side down, on a rack in a roasting pan. Bake for 1 hour. Open up the aluminum foil and turn the breast sides up. Bake uncovered for another half-hour, basting from the juices in the foil. Should the juices dry out, add a bit of red wine or bouillon.

⁊ *Serve with buttered broccoli (you won't need a starchy vegetable, as the rice stuffing will be adequate) and a medium-dry red wine.*

ROAST GOOSE WITH BURGUNDY

Although not as popular as turkey, a roast goose makes an exceptionally good main dish for holiday fare. Although the roasting time is always the same, the goose can be stuffed in many ways to give it variety (see Chapter 11 for stuffing ideas). Also it can be cooked without stuffing, as below.

Serves: 8
Time: 4 hours

12-pound goose	2 oranges (quartered)
salt and freshly ground pepper	6 thin strips salt pork
2 cloves garlic (minced)	1 cup Burgundy wine

Wash and dry goose thoroughly, inside and out. Using a sharp 2-tined fork, prick skin all over bird. Cover breast with salt and pepper and minced garlic and bring to room temperature. Preheat oven (325° F.). In cavity, place orange quarters, squeezing them slightly. Place strips of salt pork on breast of goose and put bird on rack in open roasting pan. Bake for 1 hour. Drain off all excess fat. Increase heat to 350° F. and continue roasting for 3 hours. During this cooking period, you will find that the goose has given off quite a bit of fat. The best way to dispose of it is to remove goose and place on a new rack in a new roasting pan. During the last hour of cooking you will find that most of the fat has left the bird. Now is the time to baste it with Burgundy wine. This will give a fine aroma, so baste it often, adding more Burgundy if needed.

ʆ *Serve with red cabbage, candied yams, and a Burgundy wine.*

9

Variety Meats

VARIETY MEATS INCLUDE those parts of the animal (or bird) that have not been defined in any previous description of beef, veal, lamb, pork, or fowl. Among them are great delicacies as well as cuts that are considered rather plebeian. But take a look at a menu in a French restaurant and you will see that sweetbreads, brains, and chicken livers have top billing (and top prices).

Variety meats give you a chance to use your culinary inventiveness, since their method of preparation is all-important.

⁊ LIVER

Liver has great nutritive values, especially when it is not overcooked. Doctors frequently prescribe it for invalids, even for children who are in poor health—in which case it is sometimes chopped up and cooked to a very slight degree. Liver can be cooked in butter or in bacon fat, which gives it a distinct flavor. This is a classic that restaurants specializing in hearty meats frequently tout.

There are several varieties of liver. We'll start with what we consider the finest.

CALF'S LIVER

This is the most delicate, the sweetest, and the tastiest of all animal livers. The entire liver weighs from 2 to 5 pounds.

BEEF LIVER

Even though this is a bit less tender, its nutritional value is much higher. It has a stronger and more definite taste. This liver weighs from 10 to 16 pounds and costs somewhat less than the calf variety.

LAMB LIVER

This tiny liver weighs from ¾ to 1½ pounds. It is delightful and tender but not as flavorful as calf's or beef liver.

PORK LIVER

This is sometimes used for pâté. It is similar in size and weight to lamb liver, but it is not a popular cut of meat—it has quite a strong taste.

CHICKEN LIVERS

These are exceptional in taste and flavor. But be sure to buy them from a shop that sells them fresh. Chicken livers are fresh if they have a glossy look, stand solidly, and have a firm shape.

⁊ KIDNEY

The English love beef and kidney pie, and in many English novels kidney stew is served for breakfast in those handsome Sheffield silver "pots" over hot water. In the United States, broiled kidneys are a specialty in restaurants that cater to meat-eaters. Aside from their unusually distinctive taste and flavor, kidneys are very nutritious.

VEAL KIDNEYS

These are the most delicate and sweetest of all. They are delicious when split and broiled. One of our specialties is to put a veal kidney into a rolled loin or rump of veal.

BEEF KIDNEYS

Much larger than veal kidneys, these are husky pieces of meat, and have a very strong flavor and pungent odor when cooked. They are sometimes used for stews but demand a long cooking period.

LAMB KIDNEYS

These are small, delicious, and can be cooked in many ways. As we mentioned in the lamb chapter, these kidneys may be tucked into loin lamb chops, and sometimes (for special occasions) we insert them into a rolled roast of lamb. There are many other exciting things to do with lamb kidneys. They can be slit and broiled. They can be part of a shish kebab. They can be cut up and made into a marvelous casserole with mushrooms and onions and sherry.

PORK KIDNEYS

Even though these have a strong flavor, they are delicious when tucked into a loin pork chop. They are about the same size as a lamb kidney.

⁄ TONGUE

This is a delicious and tender cut—just great served hot for a large family, marvelous cold for a buffet, or as leftovers for sandwiches. You have a wide choice when you buy tongue. The size depends on the animal it came from, and you will have a choice of fresh, pickled, or smoked.

BEEF TONGUE

This comes from the steer and usually weighs from 4 to 6 pounds. We only sell a #1-quality tongue, which is gray to pink in color and completely blemish-free. A beef tongue will serve from 6 to 10 people.

CALF'S TONGUE

This weighs only ¾ to 1½ pounds and is more tender and less fatty than beef tongue. It is excellent when potted or as a sweet-and-sour dish. A medium size will serve two people.

LAMB TONGUE

A tiny delicacy, this weighs only ¼ to ½ pound. It is always sold fresh and is used for potting.

PORK TONGUE

This is about the same size and weight as lamb tongue, but it has a much stronger flavor. It is also used for potting.

⸴ SWEETBREADS AND BRAINS

Both of these are delicate and can be cooked to make gourmet dishes. They are much alike in tenderness and texture. Parboiling is indicated, after which the thin outer layer of skin and the muscles are removed.

CALF'S SWEETBREADS AND BRAINS

These are delicate and sweet. They are, of course, smaller than those that come from a steer and cost slightly more per ounce. They are almost white in color.

BEEF SWEETBREADS AND BRAINS

These are gray to white in color and have thicker membranes. Though not as delicate as those from a calf, they are equally delicious.

⚊ HEART

This is not as popular as many other variety meats but can be a most delectable dish. The heart of the calf has a sweet, fine taste and texture. It weighs from 1½ to 2 pounds, which makes one just right for four people.

⚊ TRIPE

This is the inner stomach of an animal and may be bought fresh, pickled, or canned. The fresh variety is usually partially cooked by the butcher before he sells it. But even so, tripe takes 2 or more hours to cook.

SAUTÉED CALF'S LIVER WITH HERBS

Emily Wilkens's recipe for calf's liver emphasizes her concern with nutrition and good taste.

Serves: 4
Time: 5 minutes

2 pounds calf's liver
4 tablespoons unsaturated oil
  (combination safflower, soy,
  and peanut)
1 tablespoon parsley (chopped)

1 tablespoon chives (chopped)
powdered kelp instead of salt
  (Parkelp) or hickory smoked
  yeast (Bakon)

Ask butcher for middle cut of liver and have skin and gristle removed. Heat oil in large heavy skillet over high heat. Sauté the liver for 2½ minutes on each side and remove to a hot platter. Garnish with parsley and chives and season with Parkelp or Bakon.

/ *Serve with endive salad and a light dry Chilean wine (chilled)— and for dessert, fresh melon balls with a squeeze of lemon and a touch of honey.*

SAUTÉED CALF'S LIVER WITH ONION SAUCE

Serves: 4
Time: About 10 minutes

1½ pounds calf's liver (thinly sliced)	3 tablespoons oil
flour	3 medium onions (chopped)
salt and freshly ground pepper	½ cup bouillon

If butcher has not done it for you, remove outside skin and gristle. Now dry liver slices on paper towels and sprinkle with flour that has been seasoned with salt and pepper. Place oil in large skillet. When sizzling, add liver and sauté quickly on each side. Remove to a hot platter and add onions to remaining oil. When they are lightly browned, add bouillon and let bubble furiously. Reduce heat and simmer until sauce is somewhat reduced. Pour sauce over liver and serve.

/ *Serve with fluffy rice, buttered asparagus, and a dry red wine.*

LIVER AND BACON PUFFS

Serves: 4
Time: About 10 minutes

oil	½ pound bacon (crisply cooked)
1½ pounds calf's liver	½ cup milk
1 cup flour	½ cup water
salt	
1 teaspoon baking powder	

Preheat deep fryer (400° F.) with oil 1½ inches deep. Cut

liver into 1½-inch squares and dry them on paper towels. Meanwhile, sift flour, salt, and baking powder into a bowl. Add bacon, which you have pulverized, and stir well. Make a well in the center of this mixture and slowly add milk and water. Stir gradually, then beat well with a whisk so that the mixture becomes a smooth batter. Dip the meat pieces into this and then plunge into deep fat. Remove when they are golden brown.

⁊ *Serve with broiled tomatoes (topped with curry), a crisp endive salad, and a dry red wine.*

⁊ *Variations: This very same dish can be made with the same poundage of chicken livers or goose livers. And, instead of being a main course, it can easily be an excellent hot hors d'oeuvre, with the meat cut into smaller pieces.*

CHICKEN LIVERS WITH WINE

Serves: 2
Time: About 20 minutes

1 pound chicken livers	rosemary
2 tablespoons butter	½ cup dry white wine
salt and pepper	8 shallots (chopped)

To make this dish enjoyable, the livers must be plump and firm. Then you must dry them completely in paper towels—you may have to do this on 2 sets of towels. When they are dry enough for the paper almost to adhere to them, toss them into a heavy skillet in which the butter is sizzling. Let them brown on one side and then turn quickly (each side should take about 5 minutes). Now you can sprinkle them with salt and freshly ground pepper, and a good pinch of rosemary. When the other sides are nicely brown, stir in the wine and the chopped shallots. Turn down the heat and let everything simmer for about 10 minutes. You may have to add a bit more wine to make the dish juicier.

⁊ *This is a delicious dish for either lunch or supper. In either case, rice will absorb those nice juices—then a Chablis wine.*

CHICKEN LIVER DUMPLINGS

Serves: About 8
Time: About 20 minutes

36 chicken livers
4 tablespoons chicken fat
½ cup shallots (sliced)
2 tablespoons parsley (chopped)
16 medium potatoes (boiled and riced)

2½ cups flour
salt
¼ teaspoon baking powder
3 eggs
4 slices bacon
4 tablespoons parsley (chopped)

Drain livers on paper towels. Heat chicken fat in large frying pan and add shallots. When lightly brown, move to one side and add livers and brown on all sides. Remove livers to paper towels and add 2 tablespoons parsley to fat and shallots. Cook only 1 minute and remove from heat. Now mix riced potatoes, flour, salt, and baking powder lightly. Make well in center, drop eggs into it, and blend with hands until soft but workable. Roll dough with floured rolling pin to ¼-inch thickness, then cut into 2½-inch squares. In each square, add one well-drained liver and small amount of shallots and parsley. Fold dough so that meat is completely enclosed. In a large saucepan, bring 4 quarts of water to a rolling boil and gently immerse dumplings and reduce heat. At first, dumplings will sink to bottom, then they will rise to top. Simmer uncovered for 8 to 10 minutes. Meanwhile, fry bacon until crisp and drain on paper towels, then crumble. Add 4 tablespoons parsley to hot bacon grease and stir just long enough to moisten—but not to cook out color. Mix parsley with crumbled bacon. Lift dumplings from water gently with slotted spoon—be sure all water drains off. Garnish dumplings with a sprinkle of bacon and parsley.

ɟ *Serve with buttered string beans, endive salad, and a dry white wine.*

GIBLET FRICASSEE

There are people who feel that giblets are quite unimportant in a menu; however, there are others who find them quite a delicacy —even a gourmet dish. Giblets can be boiled for about 30 minutes with salt and perhaps a touch of oregano or rosemary. When they are drained and chilled, they can be sprinkled with paprika, sliced, and served with cocktail picks to become excellent hors d'oeuvres. But you can cook them as a main dish, such as this.

Serves: 4 to 6
Time: About 1½ hours

1 pound chicken giblets (cut up)	water
4 tablespoons chicken fat	salt and freshly ground pepper
4 medium onions (chopped)	paprika
4 stalks celery (chopped)	1 pound sirloin (chopped)
1 green pepper (diced)	½ cup seasoned breadcrumbs
	1 egg (beaten)

Drain cut-up giblets on paper towels. Heat chicken fat in large skillet and brown giblets quickly. Move to one side and add onions. When they are light golden brown, push over with giblets and add celery and green pepper. When they are limp, stir all together and add water (enough to cover bottom), salt and pepper, and paprika. Cover and simmer for about 1 hour (adding water if necessary). Meanwhile, mix chopped meat, breadcrumbs, and egg. Form this mixture into walnut-size balls. Add these to the giblets and vegetables. Cover pan and cook for about 30 minutes longer over low heat. Uncover the last 10 minutes of cooking and if sauce is too thin, mix a little arrowroot with a tablespoon of cooled sauce and add.

✓ *With such a delightful sauce, you should have a fluffy array of rice to serve it on. Then, to make the plates look more colorful, how about crisp, buttered string beans? A dry red wine would also be nice.*

BASIC FRESH BEEF TONGUE

Many people just like to have beef tongue in its good, simplified form—boiled, sliced, and served hot (without any of the frills of sauces and combinations). Here is the basic beef tongue for you.

Serves: 8 to 10
Time: 2½ to 3 hours

5- to 6-pound fresh beef
  tongue
water (boiling)

2 tablespoons salt
10 peppercorns
2 bay leaves

Place tongue into a large pot with enough boiling water to cover. Add salt, peppercorns, and bay leaves. Cover pot and cook over low heat for 2½ to 3 hours. Tongue is done when small bones at large end come out with ease. Remove tongue from water (but keep water simmering). Peel off the skin and return the tongue to hot water, just to heat through. Remove again and slice.

/ *Serve with small boiled potatoes, buttered and sprinkled with minced parsley, cabbage wedges, a sauce made with ½ pint of sour cream and 4 teaspoons of horseradish, and a good, dry red wine.*

TONGUE AND HAM WITH ORANGE

Serves: 8 to 10
Time: About 3 hours

5- to 6-pound fresh beef
  tongue (use recipe for Basic
  Fresh Beef Tongue)
¼ pound butter (melted)
1 tablespoon orange rind
  (grated)

2 tablespoons parsley (chopped)
¼ pound ham (thinly sliced,
  quartered)
2 oranges (thinly sliced)

After you have boiled tongue according to the Basic Fresh Beef Tongue recipe, slice it and spread half of the slices in a large flat baking pan. Preheat oven (275° F.). Meanwhile, combine melted butter, orange rind, and parsley. Brush this lightly over tongue slices and place a quartered slice of ham on top of each tongue round. Brush this with butter mixture and place a slice of tongue on top. Brush again with butter and lay orange slices on top. Bake for about 15 to 20 minutes.

⸎ *Serve with candied yams and broccoli with lemon butter. Beer would be nice with this.*

FRESH BEEF TONGUE WITH VEGETABLE SAUCE

Serves: 8 to 10
Time: About 3 hours

5- to 6-pound fresh beef tongue
4 strips bacon
3 large onions (thickly sliced)
3 large tomatoes (skinned, halved)
2 large potatoes (quartered)
water
salt and freshly ground pepper
2 bay leaves

Have tongue in readiness while vegetables are prepared. Place bacon strips in bottom of Dutch oven and let cook until lightly brown. Add onions, tomatoes, and potatoes and cook at medium heat for about 10 minutes. Lay tongue on top and add enough water to cover vegetables. Increase heat and cook for another 10 minutes. Add salt, pepper, bay leaves, and a little more water if it has cooked out. Cover pan and reduce heat. Simmer for about 2½ hours. Take a look now and then to see if more water is needed. You will know that tongue is done when the small bones at back end come out easily. During cooking period, turn tongue several times. When done, remove tongue from pot but keep sauce simmering. Peel off skin and slice tongue in ¾-inch pieces. Place these slices back into the simmering vegetable sauce and allow to heat through.

⸎ *This is such a hearty dish and the gravy is so tasty that crisp French or Italian bread makes a nice dipper. A tossed green salad with a tart sauce and a Burgundy wine are also good.*

LAMB KIDNEY AND MUSHROOM CASEROLE

Serves: 4
Time: About 25 minutes

12 lamb kidneys	salt
4 tablespoons butter	freshly ground pepper
1 medium onion (chopped)	thyme
½ pound mushrooms (sliced)	½ cup dry sherry

Preheat oven (450° F.). Remove gristle and cut kidneys into small pieces and drop into skillet sizzling with half quantity of butter. While they are browning, place remaining butter in another skillet. When sizzling, add onions. When they have become yellowish, push them to one side of skillet and add mushrooms. Sprinkle contents of both pans with salt and pepper and turn as they brown. When mushrooms are tender to touch of fork, pour both mixtures into deep casserole. Add pinch of thyme and sherry, cover, and cook in oven for 10 minutes.

⸴ *Tip: If your guests aren't starving, serve them another drink and turn oven down to 225° F. But first see if more sherry is needed to make casserole juicy. Extra cooking really improves the flavor of this dish.*
⸴ *Serve with steamed rice, a tart mixed green salad, and a good dry wine, either red or white.*

SWEETBREADS IN HAM CUPS

Serves: 4
Cooking Time: 5 to 8 minutes

2 large pairs sweetbreads	6 large pats of soft butter
1 tablespoon vinegar	salt and freshly ground pepper
4 slices baked ham (from	paprika
boneless round ham with fat	nutmeg
left on)	

PREPARATION: This is what takes time. Soak sweetbreads in cold water for about an hour. Drain and place them in boiling water with vinegar. Let them cook for 10 to 15 minutes. Pour off hot water and immerse them in very icy water. When they are cool enough to handle, drain and take off outer skin and connective tissues. Let them rest on paper towels.

COOKING: This is the fast part. Preheat broiler. Place ham slices in large shallow pan. Be sure that the fat is still on the edges—do not cut it off or slither it. Spread part of butter on bottom sides of each slice. Place ½ sweetbread in center of each ham slice and spread rest of butter on top. Sprinkle with salt and pepper, paprika, and nutmeg. Place under broiler for 5 minutes. This should make the ham curl up into cups and cause the sweetbreads to have a brownish look. If not, baste and add more butter and paprika.

/ *Green would make this dish pretty, so how about Italian string beans? Then a delicate dry white wine, such as the Italian Soave.*

SWEETBREADS MARINATED IN YOGURT

Serves: 4
Time: About 40 minutes

2 large sets veal sweetbreads	salt and freshly ground pepper
1 tablespoon vinegar	2 shallots (chopped)
1 pint yogurt	2 tablespoons breadcrumbs
⅛ pound butter	paprika
1 tablespoon capers	

Parboil sweetbreads in salted boiling water with vinegar for 10 to 15 minutes, then plunge them into ice-cold water. When cool, remove skin and gristle. Cut in half and cover with yogurt and let stand for 6 hours. Melt butter in large skillet and add capers, salt and pepper, and shallots. Drain sweetbreads (but do not rinse away remaining yogurt) and add to butter mixture. Brown lightly and reduce heat. Simmer for about 30 minutes covered. Add breadcrumbs and continue simmering for about 10 minutes. Before serving, sprinkle with paprika.

/ *Serve with thin buttered noodles, chopped spinach, and a chilled Chablis wine.*

CREAMED SWEETBREADS
MARYLAND NO. 1

Serves: 4
Time: About 40 minutes

2 sets veal sweetbreads	salt and freshly ground pepper
1 tablespoon vinegar	paprika
flour	rosemary
6 shallots (or 2 small white onions)	½ cup dry white wine
	1 cup cream
3 tablespoons butter	1 egg yolk

Put sweetbreads into salted boiling water with vinegar. Boil briskly for 10 to 15 minutes, then plunge into ice-cold water. When cool enough to handle, remove skin and gristle. Cut into medium-size pieces and drain well on paper towels. Place on dry towels and dredge with flour. Peel and slice shallots (or onions) and place in skillet with sizzling butter. When transparent, add sweetbreads and sprinkle with salt and pepper, paprika, and pinch of rosemary. If they should need more butter, add a bit. When all is golden brown, add wine and let simmer for about 5 minutes. Meanwhile, stir egg yolk into cream and add very slowly to sweetbreads. Keep simmering until sauce thickens, but do not boil.

ɬ *Tip: The cooking time is cut in half if your butcher has already removed the skin and gristle—you will not have to parboil at all. But the time in the skillet will take a couple of minutes longer.*

CREAMED SWEETBREADS
MARYLAND NO. 2

The recipe for No. 1 is followed exactly except that something else is added:

½ pound sliced mushrooms	1 tablespoon butter

The difference here is that you cook the mushrooms and butter in a separate pan for a few minutes (3 to 5), adding them to the sizzling sweetbreads and onions before you add the wine. Then you proceed in the exact same way.

⟡ *Either one of these two southern sweetbread dishes can be served in various ways. They can be poured into patty shells or over crisp toast with peas on the side. Either one would be excellent for supper or luncheon, but they also could be part of a late breakfast or brunch, with waffles.*

SWEETBREAD AND VEAL CASSEROLE

Serves: 8
Time: About 1 hour

1½ pounds veal sweetbreads
1 tablespoon vinegar
3½ pounds leg of veal
  (1½-inch cubes)
¼ pound butter
¾ cup green pepper (chopped)
½ cup shallots (chopped)
½ cup flour
1 teaspoon dry mustard

2 cups milk
1 small can mushroom caps
1½ cups stewed tomatoes
1 tablespoon Worcestershire
  sauce
1 cup Parmesan cheese (grated)
salt and freshly ground pepper
  to taste

Parboil sweetbreads in salted boiling water with vinegar for 10 to 15 minutes. Plunge into ice-cold water and, when cool enough to handle, remove outer skin and gristle. Drain on paper towels and cut into quarters. Meanwhile, drop veal cubes into salted boiling water and cook for about ½ to ¾ hour. Drain and cool. Place butter into large casserole dish. When bubbling, add green pepper and shallots. When they have become soft, move to one side and gradually add flour and mustard to butter. Stir constantly until thickened, then slowly pour in the milk. This should make a fairly thick sauce that will be loosened up a bit with the addition of mushroom caps and juice. Now add tomatoes (reserve juice), Worcestershire sauce, grated cheese, sweetbreads, and veal. Stir this mixture well and add salt and pepper to taste. If sauce seems too thick, add some tomato juice. Cover and simmer for about 45 minutes. Stir from time to time and add more juice as needed.

⟡ *Serve with boiled potatoes, green salad, and a dry white wine.*

BRAIN SALAD

This idea for a salad might not seem immediately appealing. However, it is quite flavorsome and is a delightful and cooling dish for a summer weekend luncheon. Even though this recipe is only for 2, it can easily be expanded. And it would make a surprise addition to any buffet luncheon party you might give.

Serves: 2
Time: About 30 minutes

1 set calf's brains
2 garlic cloves (crushed)
1 tablespoon lemon juice
½ cup salad oil
1 teaspoon vinegar
salt and freshly ground pepper
1 small cucumber (peeled
   and sliced)

3 scallions (chopped)
heart of romaine
10 cherry tomatoes
¼ cup Parmesan cheese
   (grated)
½ cup croutons (optional)

First the brains must be parboiled in salted boiling water. Then they must be placed in ice-cold water and drained. This should take about 15 minutes. When cool enough to handle, they should be skinned. Place the brains in a mixing bowl along with the crushed garlic cloves, lemon juice, oil, vinegar, and salt and pepper. Now mash this mixture so that it becomes smooth. Meanwhile, drain the cucumber slices on paper towels. Add these to a large salad bowl with the chopped scallions and the hearts of romaine, which you have either chopped or broken with your fingers. Add the brain mixture and toss well. Garnish with cherry tomatoes and cheese. Croutons may also be used as a garnish.

BROILED BRAINS AND BACON

Serves: 2 to 4
Time: About 12 minutes

2 large sets calf's brains
1 tablespoon vinegar

8 strips bacon (cut in half)

Parboil brains in salted boiling water with vinegar for 10 minutes. Preheat broiler. Plunge brains into ice-cold water. When cool, remove outer skin and veins and drain on paper towels, cutting sets in half. Place bacon-strip halves on bottom of shallow baking pan and arrange halves of brains on top of each 2 strips. Now cover each half of brain with 2 of the half-strips of bacon. Place under broiler for 6 minutes. Turn and broil for another 5 to 6 minutes or until bacon is quite crisp.

⚊ *Serve with jumbo asparagus with lemon-butter sauce and a chilled white wine such as an Italian Soave Bolla.*

CALF'S BRAINS WITH CAPERS

Serves: 4
Time: About 25 minutes

3 sets calf's brains	paprika
1 tablespoon vinegar	rosemary
flour	2 tablespoons capers
2 tablespoons butter	¼ cup dry white wine
salt and freshly ground pepper	

Place brains in salted boiling water and add vinegar. Let boil briskly for 10 minutes and then place in ice-cold water. When cool enough to handle, skin and dry between paper towels. Cut into ½-inch slices and spread on dry towel, then sprinkle with flour on all sides. Be sure they are quite dry (not gummy). When butter is sizzling in skillet, add brain slices. Brown on each side and sprinkle with salt and freshly ground pepper, paprika, then a pinch of rosemary. If butter has become absorbed, add a little more so that enough of it surrounds the slices of brains and becomes slightly brown. Now add capers with a little of their own juice, then the wine. When everything is bubbling, serve.

⚊ *Peas are obvious, but why not try artichoke hearts? And then a chilled dry white wine.*

SAUTÉED BRAINS AND MUSHROOMS

Serves: 4
Time: About 15 minutes

2 large sets calf's brains
1 tablespoon vinegar
¼ pound butter
½ pound mushrooms (sliced)
1 tablespoon capers

1 teaspoon lemon juice
2 tablespoons parsley (chopped)
salt and freshly ground pepper
dry vermouth

Place brains in salted boiling water and add vinegar. Cook briskly for about 10 minutes and immerse in ice-cold water. When cool, remove outer skin and veins, drain on paper towels, and cut in half. Melt butter in large frying pan. When bubbling, sauté brains and then remove to paper towel to drain. Add mushrooms to butter and when they have turned slightly dark and are limp, add capers, lemon juice, and parsley. Return brains to pan and season with salt and pepper. Stir well and add a bit of vermouth to loosen up sauce. Reduce heat and simmer until time to serve (adding more vermouth if necessary).

⚹ *Serve with fluffy rice, buttered spinach, and a white wine.*

VEAL HEARTS AND RED WINE CASSEROLE

Serves: 4
Time: About 35 minutes

2 veal hearts
4 tablespoons butter
8 scallions (chopped)
½ pound mushrooms (sliced)

salt and freshly ground pepper
oregano
½ cup dry red wine

Preheat oven (450° F.). Be sure that all gristle has been removed from hearts, then cut into small bite-size pieces. Have 2 skillets ready with 2 tablespoons of butter in each. Simultaneously place heart pieces in one pan and scallions in the other. When scallions are wilted, push to one side and add sliced mushrooms. When contents of both pans are slightly browned, combine them in a deep casserole. Add a bit of salt and pepper and oregano. Mix gently and add wine. Cover and cook in oven for 10 minutes. Inspect and add more wine if needed to make a juicy stew. Cook another 5 minutes and serve.

✦ *Tip: Lamb hearts may be used, but allow 1 to each person. They are tenderer and may take a bit less cooking time.*
✦ *Serve with peas and onions or carrots, and a fine claret wine.*

TRIPE CASSEROLE

Even though tripe is not a great favorite, it is sometimes considered a particularly fine dish among gourmets. Here is one that is quite delicious.

Serves: 6
Time: About 3 hours

2 pounds tripe (cut in 1½-inch squares)	4 peppercorns
	½ teaspoon thyme
2 tablespoons oil	2 bay leaves
1 medium onion (minced)	1 green pepper (diced)
2 cloves garlic (crushed)	salt and freshly ground pepper
1 14-ounce can stewed tomatoes	

Place tripe squares into salted boiling water and cook for about 2½ hours covered. Drain and cool tripe and place on paper towels. Heat oil in a heavy frying pan. When sizzling, add onions and garlic. Before they become brown, add the stewed tomatoes, peppercorns, thyme, bay leaves, and pepper. When this mixture has been well combined, reduce heat and simmer for about 30 minutes covered. Then remove bay leaves and add tripe to pan. Adjust seasoning with salt and pepper and, if sauce has diminished, add a bit of wine you will serve with meal to loosen it up. Stir until well heated.

✦ *Serve with tiny noodles, your favorite green vegetables, and a white wine.*

FRIED TRIPE

Serves: 4
Time: About 3¼ hours

2 pounds tripe (cut in strips)　　1 cup breadcrumbs
1 cup flour　　　　　　　　　　　4 tablespoons oil
salt and freshly ground pepper　　1 onion (chopped)
½ teaspoon dried sage　　　　　　Parmesan cheese (grated)
2 eggs (beaten)

Place tripe strips into salted boiling water and cook for about 3 hours covered. Drain and cool tripe and place on paper towels. Meanwhile, combine flour, salt and pepper, and sage. Roll tripe in this, then dip each piece in beaten eggs, and then into breadcrumbs. Put oil in a large frying pan over high heat. Add chopped onion and stir until golden. Push to one side and quickly sauté the tripe until beautifully brown on all sides. Before serving, sprinkle each piece of hot fried tripe with Parmesan cheese.

⨍ *Serve with buttered wide noodles, broiled tomatoes, and chilled beer.*

CHITTERLINGS

To people who are not from the Deep South or are not familiar with "soul" food, the word "chitterling" means nothing. Actually it is a part of the innards of a pig, which can be made into a gourmet-type dish.

Serves: 4
Time: 4 hours

10 pounds chitterlings　　　　　1 stalk celery
2 large onions　　　　　　　　　1 red pepper (chopped)

The chitterlings must be thoroughly washed. Some of the fat should be removed, but a little should be left on. Place these in a heavy pot or Dutch oven and add the onions, celery, and red pepper. Do not add any water, as this mixture will make its own moisture. Simmer for about 4 hours. When tender, drain and serve.

⚓ *Serve with a hot sauce—perhaps Tabasco—and then a spicy cole-slaw and well-chilled beer or ale.*

Game—Wild Birds and Animals

THESE DELICACIES used to be hard to come by and were quite expensive when available. Nowadays, specialty meat shops either have them or will secure them for you with a little advance notice. Their cost may be little more than that of a prime steak.

Game of all sorts is becoming more popular—not only in gourmet restaurants but with hostesses. There is great variety in wild birds as well as substantial weight differences so that you can plan easily for the number of guests involved. Some connoisseurs of wild birds feel they should be cooked lightly. "Just pass them through the oven," they say. But others disagree—and not surprisingly. Cooking time is really a matter of personal taste. Young, small birds are tender and delicate and need little cooking.

Every state has its own laws both for hunting and for game that is to be sold in retail stores. When you buy either wild bird or animal in New York, you will find that each (except squabs and rabbits) has a tag, which means that it has been approved by the New York State Conservation Department. This will be true in most states.

There is a firm rule against an individual hunter (even though licensed) selling any of the mallard ducks or deers he has bagged. No butcher or wholesale meat-dealer will accept such a purchase, no matter how good the game. Wild game is supplied by licensed breeders with private preserves or farms. They are given tags with registration numbers after the game has been inspected under the supervision of the Conservation Department. Birds are given a single tag. But a large animal— such as venison—must have six. One is placed on each of the four legs and two on the flank.

⸝ WILD BIRDS

These are the greatest delicacy of all—and what a choice you have in your gourmet meat market!

WILD TURKEY

Although this bird has a distinct resemblance to its domesticated relative, it does not have all of its juicy fat, since it has not been coddled and overfed. The wild turkey gets a lot of exercise. It lives on nuts, berries, and the like, and these foods give a gamy flavor. Some wild turkeys weigh as much as 14 pounds.

WILD GOOSE

Exactly like the wild turkey, the wild goose has much less fat than her domestic friends. This makes for easier cooking, as there is less fat to drain off. The meat of this bird is quite lean and has an attractive gamy flavor. A wild goose (which is a marvelous party dish) weighs from 10 to 14 pounds.

WILD DUCK

As almost every duck-hunter knows, there are at least 24 varieties available for shooting. More than half of these are diving ducks, the rest are surface feeders. The diving type have a lobed hind toe, and they sometimes feed on fish. This gives the bird a fishy flavor and makes it inedible. Of all of the varieties of wild duck, mallard is outstanding. It is a surface feeder—therefore no fish—and is raised in private preserves. This is the only type we handle, and all of our customers rave about it. Its usual weight is from 2 to 3 pounds.

PARTRIDGE

The partridge comes from the Himalayas and became a favorite with the English who were stationed in India. They were then brought to other countries, including the United States, to be bred and raised. Many game bird devotees consider the partridge the aristocrat of wild birds. An individual serving weighs only about 8 ounces, including the weight of the bones, so 2 birds are often necessary for each serving.

QUAIL

This bird weighs from 4 to 6 ounces, and it too should probably be served 2 to each guest.

GROUSE

This tasty bird is sometimes erroneously called pheasant in the Deep South and partridge in New England. In other sections of the country it is even referred to as quail. Scotch grouse is hunted in Scotland from late August to November. It is similar to the American variety but weights about half a pound while the American type is usually ¾ to 1 pound.

SQUABS

These birds do not need the usual Conservation Department inspection. They fit into the wild-bird category, but the ones you buy at a butcher shop have been scientifically raised on farms and get the same USDA inspection as poultry. Squabs are luscious little morsels. When they weigh about ¾ of a pound, they are called doves, but when they reach 1¼ pounds, they are called pigeons.

⁄ WILD ANIMALS

There are many wild animals available, but only a few that appeal to popular tastes. The others must be special-ordered, usually in large amounts. For example, reindeer from Alaska and boar from New Zealand may be ordered well in advance, and so may elk, buffalo, and bear from the western states. But few people want to buy a complete animal of that size. Smaller animals such as squirrels and muskrats can also be ordered. Venison and rabbit are just about all that our game-meat–lovers request, so we keep these on hand. Wild animals should be hung in your butcher shop for about a week before you take them home to cook.

VENISON

The deer you buy in a butcher shop always comes from a private preserve. They are kept in a natural enclosure so that they can live and eat as in their natural habitat; thus they do not become domesticated and lose none of their wild and gamy flavor. We buy the complete carcass of a deer, and these weigh from 60 to 150 pounds. The males are called bucks and are the heaviest in weight. The females are known as does. During their first year of growth, however, both bucks and does are referred to as fawns. These animals—which have never been domesticated—have been used for food for centuries. The favorite venison cuts in our shop are steaks and chops from the loin section. These are cooked in much the same way that beef steaks and other chops are cooked—by broiling. However, they are apt to take a longer time—to give extra tenderness. We buy a complete carcass because many people like the idea of braising or potting the shoulder, rump, and even the round for a delicious roast.

RABBIT

These animals do not need the usual Conservation Department inspection. They are raised commercially and are inspected by the USDA. As with deer, rabbits are kept in large natural enclosures so that they can live like wild rabbits and retain their gamy flavor. The rabbits we sell in our shop usually weigh from 2 to 3 pounds. When smaller, they can be broiled or roasted. But the best method of cooking is by braising or potting. We cut a large rabbit into sections for marinating. Even though the meat of a rabbit is almost white, it takes on a darker color when marinated and cooked.

ROAST WILD TURKEY

A wild turkey is not as fat as the domestic type, which is coddled and fed. Therefore a wild bird takes much less time in cooking. A wild turkey is a marvelous surprise as a main course for a distinguished party. It may be roasted as it is, or it may be stuffed. Should you stuff it, add about 20 minutes to the cooking time.

Serves: 8
Time: About 2½ hours

1 12-pound wild turkey	1 clove garlic (crushed)
2 tablespoons oil	½ teaspoon paprika
salt and freshly ground pepper	6 thin strips salt pork
½ teaspoon thyme	3 large green cabbage leaves

Be sure to wash the turkey inside and out and then dry it well. While it is approaching room temperature, combine oil, salt and pepper, thyme, garlic, and paprika. Rub this mixture on the outside and in the cavity. Preheat oven (350° F.). Place turkey on a rack in a baking pan and place the salt pork over the breast. Roast for half an hour and then place cabbage leaves over bacon. Roast again for 1¼ hours. Remove cabbage leaves and salt pork and continue roasting for ¾ hour. But be sure to baste from time to time with pan drippings.

⚡ *Should you have a nostalgia for the South, be sure to have a sweet-potato casserole with marshmallows browned on top, and, of course, a southern-type green vegetable like turnip greens (buttered, of course), cranberry sauce, and a sturdy red wine.*

ROASTED WILD GOOSE

What a wonderful idea for a festive menu. Even though your husband does not go out and shoot wild geese, gourmet butchers can secure them for you readily. Actually, they are much easier to roast than tame birds, as they are lean and do not need to have all of that fat removed. They have a gamy taste, usually weigh from 10 to 14 pounds, and should be able to serve 8 people.

Serves: About 8
Time: About 3½ hours

1 wild goose	2 stalks celery (chopped)
salt and freshly ground pepper	2 large onions (quartered)
stuffing (see pp. 271 to 278)	1 bay leaf
½ pound salt pork (thinly sliced)	3 tablespoons cognac
	cornstarch
4 cups bouillon	

Preheat oven (350° F.). Clean goose very well and remove any pin-feathers the butcher has forgotten. Dry well and sprinkle inside and out with salt and pepper. Add stuffing and truss. Place the goose on rack in a large baking pan and place strips of salt pork all over the breast. Pour the bouillon in bottom of pan and add celery, onions, bay leaf, and cognac. Cover pan and roast for 1½ hours, basting every once in a while. Uncover and cook for another hour. If goose seems to be a bit dry, cover again when cooking for the final hour, but be sure to baste often. Remove bird to hot platter, discard bay leaf, and add cornstarch mixed with cold water to thicken gravy.

⚡ *Serve with mashed sweet potatoes, creamed broccoli, and champagne.*

ROAST WILD DUCK NO. 1

Serves: 4
Time: About 1 hour

2 mallard ducks
salt and freshly ground pepper
1 lemon (cut in quarters)
4 small carrots (split
  lengthwise)

4 small white onions (cut in
  half)
thyme
1 cup dry red wine

Rub ducks inside and out with lemon and sprinkle well with salt and pepper. Let stand for half an hour. Preheat oven (425° F.). In each duck cavity place 2 lemon quarters, 2 split carrots, 2 halved onions, and a pinch of thyme. Place in roasting pan and add wine. Place in hot oven and baste frequently. Cooking time depends on size of ducks and rareness desired. Test with sharp-tined fork to see if blood appears. The usual time for an average-size medium-well-done duck is 1 hour.

/ *Serve with wild rice, currant jelly, mixed green salad, and claret.*

ROAST WILD DUCK NO. 2

Serves: 4
Time: About 1 hour

2 mallard ducks
1 cup port wine
1 medium onion (minced)
2 cloves garlic (minced)

salt and freshly ground pepper
2 oranges
flour (if needed)

Place ducks in a large bowl. Combine wine, onion, garlic, and salt and pepper and pour over ducks. Turn ducks from time to time so that the marinade reaches every part of them. They can stand at room temperature for an hour or more or can be refrigerated overnight. Preheat oven (425° F.). When ready to cook, drain liquid from cavity and fill each with ½ orange cut in quarters. Place in low roasting pan and cover ducks with slices from other orange. Baste with marinade and place uncovered in oven for about 25 minutes. Baste with marinade about every 10 minutes. Reduce heat to 350° F. and roast for 30 minutes, using the rest of marinade for basting. Remove ducks to hot platter, cut in half with poultry shears, and garnish with orange pieces. Meanwhile, place baking pan on top of stove over low flame. If juice has diminished, add a bit more wine. If it is quite liquid, add a little flour mixed with water.

✧ *Serve with brown rice, spiced peaches, and a dry red wine.*

BROILED QUAIL

These are delectable little birds that weigh about 6 ounces and are so tender that they take very little cooking time. In this recipe we are allowing one bird to a person. But judge your guests—they might wish to eat two. If you serve one bird to each person, be sure to have ample accompaniments.

Serves: 4
Time: 20 minutes

4 quail	thyme
3 tablespoons butter (melted)	2 tablespoons breadcrumbs
salt and freshly ground pepper	

Preheat broiler. Split the backs of the quail so that they lie flat. Wash and dry them well, then brush them on both sides with melted butter. Sprinkle them with salt and pepper and a pinch or so of thyme. Next, crush breadcrumbs with a rolling pin so that they are very fine and sprinkle them on both sides. Place birds under broiler (inside of birds up) and cook for 10 minutes. Turn them and broil for another 10 minutes. Sprinkle them with butter and serve on hot plates.

✧ *Serve with tiny buttered noodles, peas cooked with onions, tomato-watercress salad, and claret wine.*

PERDRIX AUX CHOUX (PARTRIDGE WITH CABBAGE)

Mrs. John C. Wilson spent many years in France before coming to this country. Her food tastes have a definite French accent, and she is particularly fond of recipes that stem from French households and country cooking. Here is one she particularly likes.

Serves: 4
Time: About 2 hours

2 partridges	few juniper berries
½ pound bacon	2 cloves garlic (minced)
1 medium-size cabbage	4 lumps sugar
4 large carrots (sliced	pinch nutmeg
lengthwise)	1 tablespoon lemon peel
8 small smoked sausages	(grated)
salt and freshly ground pepper	bouillon

Preheat oven (250° F.). Wash and dry the birds. Place the bacon in a large skillet and cook it until it is partially done, then remove to paper towels. In the bacon fat, brown the birds on all sides. Meanwhile, blanch the cabbage in boiling salted water for about 7 minutes. Drain and, when cool enough to handle, cut away stalk and hard inner part. Now cut the cabbage into fine slices. Place half of these slices into the bottom of a large casserole. Put the partially cooked bacon on top of this along with the carrots and sausages. Put the browned partridges on top of this and sprinkle with salt and pepper. Add the juniper berries, garlic, sugar, nutmeg, and lemon peel. Cover this mixture with the other half of sliced cabbage. Moisten with bouillon so that it comes halfway up the birds. Cover casserole and bake for about 1½ hours.

⸲ *This exceptional blending of wild bird, cabbage, and other well-cooked essences really does not need an accompaniment. Of course, rice would be excellent for those lovely juices, and a simple green salad, and Chablis (at room temperature).*

PARTRIDGE—SPANISH STYLE

Serves: 3 to 4
Time: 55 minutes

1 partridge	4 carrots (sliced lengthwise)
salt and freshly ground pepper	1 lemon (cut in eighths)
4 strips bacon	8 cloves
4 fresh grape leaves	½ cup white wine
4 tablespoons butter	16 Spanish olives (pitted)
8 small white onions	

Preheat oven (375° F.). Wash partridge inside and out and dry well. Sprinkle with salt and pepper. Place in open roasting pan and arrange bacon over breast. Then cover with grape leaves (if these are not available, use cabbage leaves). Bake in oven for 15 minutes. Reduce heat to 350° F. Place butter in pan and surround bird with onions, carrots, and lemon wedges punctured with cloves. Return to oven and bake for 40 minutes. After 20 minutes, remove leaves and baste with pan drippings, then baste again in 10 minutes. Should bacon become too brown, remove it before basting. Just before serving, add the white wine and allow it to sizzle before removing bird from oven. Place bird on a hot platter, surround it with vegetables, pour sauce over all, and garnish with Spanish olives.

⸗ *Inasmuch as vegetables are used with the roasted bird, none are needed—except, of course, saffron rice, then an excellent Spanish red wine.*

ROASTED GROUSE

Serves: 4
Time: 25 minutes

2 grouse (1 pound each)   4 tablespoons butter (melted)
2 sage leaves (optional)   salt and freshly ground pepper

Wash the birds and wipe dry inside and out. Preheat oven (450° F.). Place a sage leaf in each cavity. Rub butter and salt and pepper on inside and outside of birds. Place in a shallow pan and roast for 25 minutes. Baste twice with remaining melted butter. Serve while piping hot.

⸏ *Tip: The rich dark meat of the grouse needs only a short cooking time. Should the grouse weigh less than 1 pound, shorten the cooking period by 5 minutes.*
⸏ *Serve with savory wild rice, asparagus tips, and a Burgundy wine (either chilled or at room temperature).*

ROAST PHEASANT

A pheasant and a partridge are similar in shape and weight—either can weigh from 2¼ to 3 pounds. They are absolutely delicious but should not be overcooked. The time given here is for the larger bird, so use your discretion in cooking the bird you have on hand. Wild birds should be hung at your butcher's for two to six days.

Serves: 3 to 4
Time: About 55 minutes

1 pheasant or partridge   ½ teaspoon thyme
¼ pound butter (melted)   1 tablespoon cognac
rind from ½ lemon (chopped)   10 black grapes (seeded)

Wash the pheasant inside and out and dry well. Preheat oven (350° F.). While the bird is coming to room temperature, combine the melted butter, lemon rind, thyme, cognac, and grapes. Place this mixture in a blender and whir for a short time. If you do not have a blender, mix all ingredients well but chop the grapes very fine. Now place the bird in open pan in oven and let cook for 10 minutes. Increase heat to 375° F. Brush butter mixture inside and outside of heated bird and return to oven. Do not use all of the butter sauce, as you should use it to baste the bird several times during the next 45 minutes of cooking.

⸗ *Serve with wild rice (obviously), baked acorn squash, and a Burgundy wine.*

FAGIANO ALLA CREMA (PHEASANT IN CREAM)

Vieri Traxler, the Consul General of Italy for New York, comes from Tuscany, and this recipe derives from that region.

Serves: 3 to 4
Time: About 1 hour

1 pheasant	1 wine glass of cognac
butter (about ¼ pound or less)	1 cup heavy cream
	lemon juice (few drops)
2 onions (juice only)	1 tablespoon parsley (chopped)

Be sure pheasant has been hung for two to six days, then plucked and cleaned. Place butter in a flameproof earthenware casserole. When it has melted, add the pheasant and brown it on all sides (about 15 minutes). Add the onion juice, reduce heat, cover, and cook slowly for about 30 minutes. Now uncover and add the cognac and allow it to evaporate. Then add the cream. Reduce heat to almost nothing and cook for another 15 minutes. Before serving in the same casserole, add a few drops of lemon juice and garnish with chopped parsley.

⸗ *Although the Consul General did not say so, we suggest rice,* arugula *salad, and a good Italian red wine.*

ROAST SQUAB NO. 1

Serves: 2
Time: About 35 minutes

2 squabs	rosemary
½ lemon	thyme
salt and pepper	¼ pound sliced mushrooms
4 small white onions	the squabs' giblets (chopped)
1 cup dry white wine	

Preheat oven (450° F.). Rub squabs with lemon inside and out and sprinkle with salt and freshly ground pepper. Inside each squab, place 2 onions, peeled and cut in half. Place birds in uncovered casserole, adding wine and pinches of herbs. Cook for 15 minutes. Reduce heat to 350° F. and add mushrooms and giblets. If sauce has dried out, add more wine. Bottom of casserole should be covered so that giblets and mushrooms remain moist while cooking. Place cover on casserole and cook for about 20 minutes more. Of course, cooking time depends on tenderness and size of birds. A quick plunge of a sharp fork will give you the indication of tenderness.

⌐ Serve with wild rice, mixed green salad, and a dry white wine.

ROAST SQUAB NO. 2

Serves: 4
Time: About 1 hour or more

4 squabs	stuffing (see pp. 271 to 278)
1 lemon (quartered)	2 tablespoons soft butter
salt and freshly ground pepper	

Preheat oven (350° F.). Rub squabs inside and out with lemon and sprinkle them well with salt and pepper. Now fill each cavity with stuffing and truss them. Place the birds in a casserole and smooth the softened butter over them. Cover and roast for about 45 minutes. Baste them from time to time with pan drippings. Now uncover and bake for about 15 minutes.

⸏ *Note: The time difference between No. 1 and No. 2 is the difference in oven heat. Also, a stuffed bird should take at least 15 minutes more than an unstuffed one.*
⸏ *Due to the stuffing, you won't need any starches with this. So just serve with green peas and onions and perhaps a green salad. Then a rosé wine would be excellent—chilled, of course.*

PIGEON (OR SQUAB) CASSEROLE

Serves: 4
Time: About 1¼ hours

1 lemon (quartered)	¼ pound butter
4 young fat pigeons	1½ tablespoons flour
salt and freshly ground pepper	1 cup bouillon
8 dried prunes	½ pound mushrooms (sliced)
8 apricots (dried)	1½ tablespoons seedless raisins
4 truffles (quartered)	2 tablespoons parsley (chopped)

Use lemon quarters to rub the squabs inside and out. When they have come to room temperature, season them with salt and pepper. In the cavity of each, place 2 prunes, 2 apricots, and 1 quartered truffle. Truss the birds. Preheat oven (350° F.). In a large frying pan, melt the butter and brown the squabs lightly on each side. Place the birds in a large casserole. Now add the flour gradually to the butter in the frying pan. When this is thickened and not lumpy, slowly add the bouillon. Stir constantly until the consistency of heavy cream. Pour the sauce over the birds, cover the casserole, and bake for about half an hour. Add mushrooms and raisins and stir. Return to oven uncovered and bake for ¾ hour. Should sauce diminish during this last cooking period, add a bit more bouillon or a dash of dry white wine. Before serving, garnish with parsley.

⸏ *Serve with glazed carrots, a crisp tossed salad, and a rosé wine.*

BRAISED VENISON

Libby Hillman coordinates and teaches classes on adult cookery and has written several cookbooks. She has given us one of her favorite recipes—never included in any of her books.

Serves: 8 to 10
Time: About 2½ hours

4 pounds venison (shoulder or rump)
2 cups red wine
½ cup vinegar
½ cup water
1 onion
4 cloves
1 carrot (sliced)
2 bay leaves
12 peppercorns
1 tablespoon salt
½ pound bacon
2 onions (diced)
3 ribs celery (diced)
2 carrots (diced)
4 cloves garlic (whole)
3 sprigs parsley
½ teaspoon thyme
2 teaspoons salt
½ teaspoon pepper
4 tablespoons flour
2 cups beef broth
¼ cup currant jelly
2 slices rye bread (with caraway seeds)
¼ cup parsley (minced)

While venison is coming to room temperature, mix the marinade—the next 9 ingredients from wine through salt. Pour this over the venison, covering the meat completely. This meat can be marinated from three to five days, covered, in a cool place. It does not have to be refrigerated. When ready to cook, drain meat thoroughly and reserve 1 cup of the marinade. Dry the meat thoroughly with paper towels. Meanwhile, place bacon in Dutch oven. When it has crisped, remove, crumble, and reserve. Pour off all but enough bacon fat to coat the bottom of pan. Now brown venison on all sides in the remaining bacon fat. Be patient for meat to brown well—the flavor will improve if meat is well browned. Push meat to one side and add onions, celery, carrots, garlic, parsley, thyme, and salt and pepper. Sprinkle the flour over vegetables when they have become limp and stir until well blended. Cover the Dutch oven and cook for 30 minutes. In a saucepan, heat the marinade you have saved and the beef broth. Add this to the meat and continue cooking in the covered Dutch oven for 2 hours. Then add the currant jelly and the bread, which you have crumbled. Taste and adjust seasoning. Remove meat and slice. Strain the sauce and then reheat before pouring it over the meat. Sprinkle with reserved crumbled bacon and parsley.

⸭ *Serve with chestnuts in butter sauce, prunes pitted and marinated in port wine, noodles decorated with currants soaked in Madeira wine.*

VENISON GOULASH

The cooking time for this recipe is for venison that has been hung for twelve days, which makes venison delicious and adds tenderness. If it is hung for a lesser period, it should be cooked longer.

Serves: 10
Time: About 2 hours

6 pounds venison (from round)	2 large onions (chopped)
2 cups wine vinegar (garlic-flavored)	2 cups dry red wine
½ cup bacon (cooked and crumbled)	2 cups bouillon
5 tablespoons oil	thyme (several pinches)
1½ cups carrots (sliced)	salt and freshly ground pepper
1½ cups celery (diced)	¾ pound fresh mushrooms (sliced)
	flour
	1½ tablespoons chives (minced)

Have the venison cut into 1½-inch cubes. Place these in a large bowl, add the wine vinegar and bacon, and refrigerate for about 24 hours. Drain the cubes, dry, and reserve the marinade. Preheat oven (325° F.). Heat the oil in a large skillet and brown the meat on all sides over medium-high heat. Place the venison in a large casserole and add the carrots, celery, and onions. Pour the wine and bouillon over this and season with thyme and salt and pepper. Cover and cook in oven for 1½ hours. Now add the mushrooms and continue cooking for about half an hour. If sauce is too juicy, mix a little flour with water and add. Just before serving, stir in the chopped chives.

⸭ *Serve with wide noodles, baked apples, and a Burgundy wine.*

BROILED VENISON CHOPS

Serves: 4
Time: 30 minutes

4 venison chops (1½ inches thick)
2 cups milk
1 tablespoon dry mustard
salt and freshly ground pepper

thyme
½ teaspoon dry mustard
½ teaspoon lemon juice
1 tablespoon oil

Wash chops and dry them well. Meanwhile, mix the milk and 1 tablespoon of mustard well. Pour the milk mixture over the chops and allow to marinate overnight in the refrigerator. Next day, discard the milk and mustard marinade and dry the chops on paper towels. Preheat broiler. Now mix the salt and pepper, thyme, and ½ teaspoon mustard with the lemon juice and oil. Rub this mixture all over the chops and place under broiler. Cook for 15 minutes on each side.

/ Serve with stuffed potatoes with cheese topping, broiled tomatoes, and a husky red wine.

RABBIT CASSEROLE NO. 1

Serves: 4
Time: About 1½ hours

2-pound rabbit (cut in pieces)
1 cup vinegar
flour, salt and pepper
6 strips bacon
1 large onion (sliced)

oregano
basil
1 bay leaf
2 cups dry red wine

Cover rabbit pieces with water. Add vinegar and soak for an hour or longer. Remove and dry each piece thoroughly. Shake each piece in bag with flour and salt and pepper. Preheat oven (350° F.). Meanwhile, fry bacon in large skillet. Remove, drain on paper towels, and crumble into small pieces. Brown rabbit pieces on all sides in hot bacon fat. Remove to paper towels and add onions to remaining bacon fat. Cook until transparent. Arrange rabbit in casserole and add onions and bacon. Sprinkle with a good pinch of oregano, basil, and 1 bay leaf. Pour wine over all and cook for 1¼ hours, adding more wine if needed. Discard bay leaf.

⸱ *Serve with wild rice, green salad, and claret wine.*

RABBIT CASSEROLE NO. 2

Serves: 8
Time: About 1½ hours

2 2½-pound rabbits (cut in pieces)	salt and freshly ground pepper
2 tablespoons salt	¼ pound butter
½ cup flour	2 cloves garlic (chopped)
½ cup cornmeal (yellow)	2 medium onions (chopped)
½ cup light sweet cream	1½ cups cider
2 eggs	½ cup dry red wine

Cover rabbit pieces with water and add 1 tablespoon of the salt. Soak for about half an hour. Discard water and cover rabbit with fresh water and another tablespoon of salt. Soak for another half-hour. Discard water and wipe rabbit dry. Mix the flour and cornmeal. In a bowl, beat cream and eggs lightly. Dip each piece of rabbit into egg mixture, then into flour and cornmeal, and back into egg and cream. Season each piece with salt and pepper. Melt the butter in a very large skillet over medium heat. Add garlic and onion and cook until golden. Remove and reserve this, then add the rabbit pieces. Increase heat and brown on all sides. Reduce heat, return onion and garlic, and add cider. Cover and simmer for about half an hour. Add wine and stir lightly. Cover again and simmer for about 40 minutes. Uncover, increase heat, and cook until sauce diminishes.

⸱ *Serve with red cabbage, apple sauce, and Burgundy.*

RABBIT CASSEROLE NO. 3

Serves: 4
Time: 1½ hours

1 rabbit (cut in pieces)	1 small green pepper (sliced)
1 cup vinegar	2 stalks celery (chopped)
salt and freshly ground pepper	basil
2 tablespoons butter	1 bay leaf
2 tablespoons oil	10 cloves
1 clove garlic (crushed)	1 cup tomato sauce
5 scallions (chopped)	1 cup sherry wine

Soak the rabbit pieces in water to cover and vinegar for an hour (this can also be kept in the refrigerator overnight). Discard the water and vinegar and dry the rabbit. Sprinkle each piece with salt and pepper. Heat butter and oil in large frying pan over high heat. Brown the rabbit on all sides. Reduce heat and place the garlic, scallions, green pepper, and celery on top. Sprinkle with basil and add the bay leaf and cloves, which you have put in a cheesecloth bag. Simmer covered for a few minutes while combining the tomato sauce with the wine. Add this and continue simmering covered for an hour or more. Should the sauce become too thick, add a bit of water. Discard the cheesecloth bag.

⸙ *Serve with thin noodles, broccoli, and a rich red wine.*

Poultry and Meat Stuffings

HERE ARE A NUMBER of recipes that are exceptionally good both for poultry and meat. The ingredients of the stuffing you select will help you decide what else to serve with the dinner. The following are a few important tips about stuffings:

- Make the stuffing just before roasting the bird.
- Should you decide to make the stuffing the day before, be sure to keep it in the refrigerator.
- Should you refrigerate the stuffing, be sure to bring it to room temperature before stuffing.
- A stuffed bird will take from 15 to 25 minutes more to roast than an unstuffed one.
- Always remove all stuffing from leftover meat and poultry and refrigerate separately.

SAGE AND ONION STUFFING FOR POULTRY

6 medium onions (chopped)	1 tablespoon sage
4 tablespoons butter	salt and freshly ground pepper
½ pound sausage meat	½ cup bouillon
6 cups breadcrumbs	

Place onions in large skillet sizzling with butter. When golden, push to one side and add sausage meat. Reduce heat and cook only slightly. Meanwhile, season breadcrumbs with sage and salt and pepper. Add this mixture to skillet and mix well with sausage and onions. Remove from heat and add bouillon. If ½ cup does not moisten dressing the way you like it, add a bit more. Adjust seasoning.

⚹ *This is especially fine with goose (because of the sage and sausage) but is also tasty with turkey. This recipe is enough for a small turkey or goose.*

BREAD STUFFING FOR POULTRY

5 cups breadcrumbs
1 teaspoon poultry seasoning
salt and freshly ground pepper
½ cup butter

1 large onion (chopped)
½ cup celery (chopped)
¼ cup parsley (chopped)

Place breadcrumbs in bowl and add poultry seasoning and salt and pepper. Mix well. Place butter in large skillet and sauté onions and celery until soft. Remove from heat and stir in the breadcrumb mixture and parsley. If the dressing does not seem quite moist enough, add a little melted butter.

⚹ *Use with roast chicken, capon, or small turkey. For a large turkey, increase ingredients.*

WILD RICE STUFFING WITH GRAPES FOR POULTRY

1 cup wild rice
3 cups salted water
4 tablespoons butter
1 onion (chopped)
1 large stalk celery (chopped)

1 cup grapes (seeded and halved)
½ cup chopped almonds
1 teaspoon thyme
salt and freshly ground pepper
½ cup sherry wine

Wash the rice thoroughly in a sieve and drain. Place rice in saucepan and cover with salted water. Bring to a boil and cook for 15 to 20 minutes, then drain and return to pan and place over lowest possible heat so that the grains can become separated. Meanwhile, melt butter in large skillet. When bubbling, add onion and celery. When these are transparent, add the grapes, almonds, and thyme. Reduce heat and simmer for about 5 minutes. Remove from heat and add the rice. Stir well and season to taste with salt and pepper. Now add the sherry wine and mix thoroughly.

⸕ *This recipe was devised to stuff 2 guinea hens; however, it would be delightful in Cornish game hens or in any other type of domestic or wild bird. Adjust the quantities for a large bird, and should you have any left over from the small-bird stuffing, save it for a future meal.*

SAVORY WILD RICE STUFFING FOR POULTRY

This is an elegant type of stuffing for any kind of bird, especially the wild variety. Because the price of wild rice is rather dear, this stuffing will be used for parties or impressive occasions. Should this recipe be used for small wild birds, be sure to save any leftovers in the refrigerator for a future meal, as it makes an excellent side dish in place of potatoes.

2 cups wild rice	1 large onion (finely chopped)
6 cups salted water	1 clove garlic (crushed)
3 tablespoons butter	1 teaspoon thyme
2 stalks celery (chopped)	salt and freshly ground pepper

Wash rice in a sieve and drain. Place in large pot and cover with cold salted water. Bring to a boil and cook for 15 to 20 minutes or until tender. Meanwhile, melt butter in large skillet and add celery, onion, and garlic. Stir until ingredients are transparent and remove from heat. Drain the cooked rice and add to skillet mixture. Season with thyme and salt and pepper and stir well until the flavoring is adequate.

⸕ *This should be enough stuffing for a 5- to 6-pound bird, and, of course, it can be used in smaller wild birds or Rock Cornish game hens.*

SAVORY KASHA STUFFING
FOR POULTRY

Kasha is frequently called buckwheat groats. It is made from buckwheat grain and then roasted, which gives it a delicious nut-like flavor. Aside from being a tasty stuffing for poultry, this recipe makes an excellent side dish in place of rice, potatoes, or noodles. So if you have any stuffing left over, be sure to refrigerate it and save for a future meal.

1 cup buckwheat groats
1 egg (slightly beaten)
2 cups boiling water
3 strips bacon (cut in pieces)
4 tablespoons butter
1 medium onion (chopped)
1 clove garlic (minced)

½ green pepper (chopped)
¼ pound mushrooms
 (chopped)
1 teaspoon oregano
½ teaspoon sage
salt and freshly ground pepper

Mix the groats with beaten egg, then add to frying pan over high heat. Stir constantly until grains separate. Now add the briskly boiling water, cover pan, reduce heat, and simmer for 30 minutes. Meanwhile, fry bacon in another large frying pan. When bacon is lightly browned, push to one side and add butter. Let this sizzle and add onion, garlic, green pepper, and mushrooms. Stir well while they are cooking so that they do not become too brown. Add oregano, sage, and salt and pepper. Reduce heat and add the cooked groats. Mix well, adjust seasoning, and remove from heat.

ｨ *This is enough for a small roast chicken and more than enough for 2 Cornish game hens. Increase ingredients for goose.*

CHESTNUT STUFFING FOR POULTRY

16 large chestnuts
½ cup butter
1 large onion (chopped)
2 stalks celery (chopped)
8 chicken livers
5 cups breadcrumbs

salt and freshly ground pepper
½ teaspoon rosemary
¼ cup heavy sweet cream
½ cup consommé
½ cup watercress (chopped)

Split the skins of chestnuts and place them in boiling water until skins open up. Drain and peel them, then chop them coarsely. Place butter in a large skillet and, when sizzling, add the onion and celery. When these are golden, push to one side of skillet and add the chicken livers. Brown them slightly on each side and remove from heat. When they have cooled, cut them in pieces. Meanwhile, season the breadcrumbs with salt, pepper, and rosemary. Mix well and add the cream, consommé, and chestnuts. Combine this mixture with the chicken livers and onions in frying pan. Add the watercress and mix well. Test and adjust seasoning if necessary.

⁊ *This stuffing was devised especially for Boneless Capon with Chestnut Stuffing (pp. 211–212), but it is equally delicious with chicken or a small turkey.*

CHESTNUT AND BRAZIL NUT STUFFING FOR POULTRY

2 pounds chestnuts
1 pound Brazil nuts
¾ pound butter
2 small onions (chopped)

2 stalks celery (chopped)
18 slices stale bread (cubed)
salt and freshly ground pepper
1 cup heavy sweet cream

Split the skins of chestnuts and place in boiling water until skins open up. Drain and take off the skins. Place chestnuts and Brazil nuts in blender until they are well chopped (it is best to do only a few at a time). Place butter in a large skillet and sauté the onions and celery until soft. Now add the bread cubes and salt and pepper to taste. Stir well and gradually add the cream. Remove from heat and add the nuts.

⁊ *This is an excellent stuffing for turkey, but it can be equally fine for chicken or capon. If there is any left over, place in a small baking pan, refrigerate, and bake for last half-hour of roasting time.*

DATE AND NUT STUFFING
FOR POULTRY

4 slices bacon (diced)
2 tablespoons butter
1 medium onion (chopped)
4 chicken livers
giblets from birds to be
  roasted
½ cup dates (pitted and
  diced)

¼ cup macadamia nuts
  (chopped)
¼ cup parsley (chopped)
1 cup soft breadcrumbs
salt and freshly ground pepper
¼ cup sherry

Place bacon in large skillet and sauté until lightly brown. Push to one side and add butter. When sizzling, add onion, chicken livers, and giblets. When browned slightly, remove from heat. Meanwhile, mix dates, nuts, parsley, and breadcrumbs and season with salt and pepper. Combine this mixture with ingredients of skillet. Stir well and adjust seasoning, then add the sherry. This should be of the right consistency for stuffing. If you want it softer, add a bit more sherry.

/ This stuffing was especially devised for squabs and will adequately fill 4 of them. It is equally delightful for Rock Cornish game hens. Should you wish to use it for a larger bird, just add to the ingredients.

FRUIT, NUT, AND SAUSAGE STUFFING
FOR POULTRY

6 dried apricots
½ cup seedless raisins
4 dried prunes (pitted)
8 slices bacon
1 large onion (chopped)
½ pound sausage meat
5 cups stale bread (cubed)

salt and freshly ground pepper
1 teaspoon sage (optional)
16 chestnuts (cooked and
  chopped)
rind of one lemon (chopped)
dry white wine (½ cup or
  more)

Place apricots, raisins, and prunes in boiling water and parboil for about 5 minutes. Drain and chop up the apricots and prunes. Place bacon in large skillet and cook until crisp. Remove to paper towels to drain, then crumble. In bacon fat, sauté the onions quickly. Remove onion and give the sausage a quick and light browning. Remove this too and drain on paper towels. Discard the bacon fat. Meanwhile, season the bread cubes with salt and pepper and sage. Add the chopped chestnuts, lemon rind, crumbled bacon, and finally the chopped fruit. Stir well and moisten with wine until the stuffing is of a good enough consistency to stuff the bird.

⸸ *This is a marvelous stuffing not only for a goose or large duck but also for a small turkey.*

GIBLET STUFFING FOR POULTRY

1½ pounds giblets (with necks)
¼ pound butter
2 medium onions (chopped coarsely)
½ green pepper (chopped)
1 stalk celery (chopped)

4 cups breadcrumbs or bread cubes
¾ cup milk
salt and freshly ground pepper
thyme (optional)

Place giblets and necks in salted boiling water to cover. Boil for about 30 minutes or until tender. Drain, then, when cool enough to handle, strip meat from necks and chop in small pieces, cutting up the gizzards, hearts, and livers. Melt butter in large skillet and sauté the onions, pepper, and celery. Cook until they are limp but not brown and remove from heat. Meanwhile, mix the breadcrumbs with milk and season to taste with salt and pepper and thyme. Add this mixture and the chopped giblets to the skillet and mix thoroughly.

⸸ *This is enough stuffing for a good-sized turkey (12 to 14 pounds). For smaller birds, decrease that ingredients. Any giblets are excellent for this stuffing, but if you are a stickler about cooking ingredients, use the giblets of the particular bird you are roasting, whether it be turkey, goose, chicken, capon, or duck.*

OYSTER STUFFING—BALTIMORE STYLE—FOR POULTRY

1½ pints oysters
4 tablespoons butter
1 large onion (chopped)
1 half-cup celery (chopped)

6 chicken livers
5 cups stale bread (cubed)
salt and freshly ground pepper
rosemary

Simmer the oysters in their own juice for about 3 minutes. Remove from heat. Drain the oysters (reserving the broth) and cut them in small pieces. Place the butter in a large skillet and sauté the onion and celery. Move them to one side of skillet and sauté the chicken livers. Brown them lightly on each side and remove from heat. Cut the livers in small pieces. Meanwhile, season the bread cubes with salt and pepper and rosemary to taste. Now add the bread cubes to the skillet and mix well with the onion, celery, and livers. Add the oysters. Adjust seasoning and add as much oyster broth as you feel will be needed to make the type of moist dressing you like.

⸱ *This is a favorite with Maryland people for their Christmas-dinner turkey. It also tastes great with a goose and is enough for an 8-pounder.*

POTATO STUFFING FOR POULTRY

2 cups mashed potatoes
4 eggs (beaten)
2 cups stale bread (cubed)
½ cup chicken broth

4 tablespoons chicken fat
1 large onion (chopped)
salt and freshly ground pepper
 to taste

Combine potatoes with eggs. In another bowl, mix bread cubes with chicken broth. Combine these 2 mixtures. Meanwhile, heat chicken fat in a large skillet and sauté the onion. When it is golden but not brown, combine all ingredients and season to taste with salt and pepper.

⸱ *Use with roast chicken, capon, or small turkey. For a large turkey, increase ingredients.*

MEAT STUFFING FOR CROWN
OF LAMB

1 pound lamb (ground)
¼ pound sirloin (ground)
1 tablespoon chicken fat

1 clove garlic (minced)
1 teaspoon thyme
salt and freshly ground pepper

Mix the lamb and sirloin in a large bowl. Meanwhile, place chicken fat in a large skillet. Add the garlic and cook until transparent. Sprinkle this with thyme and salt and pepper. Reduce heat and add the meat mixture. Stir well and test for seasoning. Remove from heat and let cool.

⁊ *Tip: Before stuffing, rub inside of crown with lemon juice, or rub inside with fresh mint leaves.*
⁊ *Obviously this is a stuffing that is only for a handsome crown of lamb, but it can be used on its own for delicious lamb-sirloin patties or to stuff red or green peppers.*

FRUIT STUFFING FOR CROWN LOIN
OF PORK

This stuffing would also be excellent for a duck, a goose, or a wild goose.

1 cup butter
2 large onions (chopped)
20 slices bread (broken into pieces)
1 cup apricots (dried)
1 cup prunes (dried)

6 apples (peeled and cored)
salt
1 teaspoon nutmeg (scant)
1 teaspoon cinnamon (scant)
½ teaspoon cloves (ground)

Melt the butter in a large pan. When it is bubbling, add the onions and allow to cook gently until they are transparent. Add the bread pieces to this mixture and remove from heat—after you have stirred well. Meanwhile, parboil the apricots and prunes for about 4 minutes. Drain them and cut into small pieces. Add these to the large pan along with the apples you have chopped into medium-size cubes. Stir this mixture thoroughly and add the seasonings. Should you feel that the mixture has not melded well, add a little water, wine, or bouillon. When the mixture seems just right, test it for seasoning and add a bit of this or that to bring it to your own taste.

MEAT AND SAUSAGE STUFFING
FOR CROWN OF PORK

A crown of pork is a handsome dish for a large party, especially if it is the complete loin. This stuffing was planned for just such an impressive occasion.

1 pound pork (ground)	1 onion (chopped)
½ pound sausage meat	1 apple (skinned and chopped)
8 dried prunes	½ teaspoon sage
8 dried apricots	salt and freshly ground pepper
1 tablespoon butter	2½ cups croutons
2 shallots (chopped)	apple juice

In a large bowl, mix the ground pork and the sausage meat. Meanwhile, parboil the prunes and apricots for about 5 minutes. Drain them and chop into small pieces (taking out prune pits). Now melt butter in a large skillet and, when bubbling, add the shallots, onion, and then the apple. Season this with sage and salt and pepper. Stir well, then add the pork and sausage mixture. Stir the meat so that it will become slightly browned. Add the fruit and adjust seasoning. Reduce heat and add the croutons, then the apple juice (the amount of juice depends on how much moistening the stuffing needs—from ½ to 1 cup).

⸹ *Although it was designed for a crown of pork, this stuffing can be used for pork patties and for stuffing large lean birds.*

STUFFING FOR BEEF SCALLOPINI ROULADES

Although this delicious stuffing was devised especially for Beef Scallopini Roulades (pp. 52–53), it is such a delight that people use it in many ways (see below).

4 tablespoons butter
1 large onion (finely chopped)
6 large mushrooms (chopped)
2 cloves garlic (crushed)
½ teaspoon thyme
caraway seed (about 2 pinches)
celery seed (about 4 pinches)
salt and freshly ground pepper
1 tablespoon parsley (chopped)
1 cup breadcrumbs
1 egg (beaten)
3 tablespoons heavy cream

Place the butter in a large skillet and let it bubble. Add the onion, mushrooms, and garlic and allow them to become wilted. Then reduce the heat and add the thyme, caraway seed, celery seed, salt and pepper, and parsley. Meanwhile, combine breadcrumbs, egg, and cream. Add this mixture to skillet and remove from heat. Mix well and test for proper seasoning.

⁊ *Aside from Beef Scallopini Roulades, this stuffing goes well with veal birds or other meat dishes that require a bit of stuffing. It can also be used with any type of bird.*

Sauces

BÉCHAMEL SAUCE

Although originated in France, this sauce has become a basic for American cooks, who refer to it merely as "white sauce." It is very easy to make, as it involves only a roux (equal amounts of butter and flour mixed well) with the addition of very hot milk. This sauce is most versatile. To the basic ingredients, all sorts of interesting diversions can be added. For example, parsley and/or herbs may be introduced, or dry mustard can be added to the flour for a sharp taste. And then, the sauce could readily become mornay sauce by the simple addition of grated cheese (almost a cup, with a slight seasoning of Worcestershire sauce and mustard).

2 tablespoons butter   1 cup milk
2 tablespoons flour   salt and white pepper

Melt the butter in a skillet over medium heat. Gradually add the flour. Should the flour seem to become a bit brownish, remove from heat and continue stirring. Meanwhile, heat the milk in a double boiler. It should become very hot—but not to the boiling point. Return the skillet to the heat and pour the milk in, stirring continuously. Now add the salt and pepper (be sure to use white pepper, as the sauce should not have any little black spots in it). When the flavor appeals to you, reduce heat to almost nothing and serve hot over any dish you have ready.

HOLLANDAISE SAUCE

¼ pound butter	1 tablespoon lemon juice
3 egg yolks	salt and cayenne pepper
1 tablespoon water	(to taste)

Divide butter into thirds and soak in ice-cold water. Beat egg yolks, cold water, and lemon juice in top of double boiler. Place this over very hot (not simmering or boiling) water. Continue beating with a wire whisk and add ⅓ of the butter. When it melts, add another third—but continue whisking. Then add the last third. When this has melted and blended well, remove from the hot water and add seasonings. Should the sauce seem too thick for your taste, thin it with a bit of hot water. This sauce is actually quite easy to make, but it involves your undivided attention and a constant whipping with a whisk. This recipe will make about a cup of sauce—enough to embellish cauliflower or asparagus (for 8 or more servings) and eggs Benedict (4 servings).

CREAMY MUSTARD SAUCE FOR HAM

¼ pound butter	¼ cup cider
¼ cup dry mustard	1½ cups beef broth
½ cup sugar	4 egg yolks (beaten)
1 tablespoon port wine	4 tablespoons heavy sweet
¼ cup vinegar	cream

Melt the butter in a saucepan and gradually add the mustard and sugar. When this becomes a paste, stir in the wine, vinegar, and cider. Now add 1 cup of the beef broth. Meanwhile, mix the other ½ cup of broth with the beaten egg yolks. Add this mixture very slowly and stir constantly over low heat so that it does not thicken too fast. When the sauce has reached the consistency you desire, remove from heat and stir in the heavy cream. The sauce should be warm enough, but should there be a waiting period, place it in a double boiler to keep hot.

BÉARNAISE SAUCE

2 tablespoons shallots (minced)
1 tablespoon tarragon vinegar
½ cup white wine
5 egg yolks
1½ teaspoons tarragon
3 peppercorns
salt
½ teaspoon dry mustard
¾ cup butter (melted)
pinch cayenne pepper
squeeze of lemon juice
  (optional)

Simmer the shallots in vinegar and white wine until liquid has been reduced by about ⅔. Remove from heat and cool. Place this mixture in blender and add egg yolks, tarragon, peppercorns, salt, and dry mustard. Blend at high speed for no more than 10 seconds. Add the warm melted butter, cayenne, and a squeeze of lemon juice (optional). Quickly turn on blender again and let it rotate until sauce thickens. Should it become too thick, add a bit of hot water and blend again. This sauce should be made at the last minute so that it will be warm when served (but it may be kept warm in a double boiler).

⸱ *Use with grilled or sautéed meat. It is quite a rich sauce, so don't plan too hearty a menu to serve with it.*

BORDELAISE SAUCE NO. 1

3 tablespoons butter
3 shallots (minced)
½ cup mushrooms (sliced)
½ cup dry red wine
1 cup beef gravy
2 tablespoons parsley (minced)
salt and freshly ground pepper
dash cayenne pepper

Melt the butter in a skillet. When bubbling, add the shallots. Cook until transparent and add the mushrooms, cooking until soft. Add the red wine and simmer uncovered until it has reduced to half its volume. Add the beef gravy and parsley and stir well. Season to taste with salt and pepper and cayenne.

⸱ *Use with grilled meats, especially steak. The classic French way to serve grilled meat and bordelaise sauce is to use a garnish of round slices of beef marrow that has been poached and drained.*

BORDELAISE SAUCE NO. 2

3 tablespoons butter
1 onion (chopped)
2 carrots (sliced in rounds)
2 cloves garlic (cut in halves)
4 peppercorns
1 bay leaf

1¼ cups beef broth
thyme
salt
2 tablespoons flour
½ cup dry red wine
1 tablespoon chopped parsley

Melt the butter in a heavy skillet and sauté the onions until transparent. Add the carrots and cook until soft. Place garlic, peppercorns, and bay leaf in cheesecloth bag. Add to pan along with 1 cup of beef broth. Boil until broth has been slightly reduced—about 10 minutes—then season with thyme and salt. Remove cheesecloth bag and put mixture through a coarse strainer. Return to pan. Meanwhile, dissolve flour in the other ¼ cup of broth and add gradually to sauce. Stir constantly until sauce thickens, then add the wine. Reduce heat and allow to simmer until ready to serve, then sprinkle chopped parsley on top.

/ *This recipe was especially devised to serve with Baked Fillet of Beef (p. 51) but is equally delicious with broiled steaks.*

BEURRE NOIR SAUCE

½ cup butter
2 tablespoons parsley (chopped)

1 tablespoon capers
1 tablespoon vinegar

Place the butter in a small frying pan and allow it to simmer until dark brown—not black (even though the name is *beurre noir*). Add the parsley and capers, stirring constantly, then add the vinegar. Instead of the vinegar, you may use the juice from the capers.

/ *Use with sautéed brains.*

⸗ BUTTER MIXTURES

Butter mixtures make excellent last-minute additions to grilled or sautéed meats. They may also be added to sauces for additional flavoring. Kneaded Butter makes a quick last-minute thickener for sauces or gravies. The virtue of these butter mixtures is that they can be made in advance and refrigerated, or even frozen. They all involve softened (not melted) fresh butter mixed with one or more ingredients.

CHIVE BUTTER

½ cup butter (softened)   salt and white pepper
3 tablespoons chives (chopped)

Cream the butter with a fork until fluffy and almost white. Mix in the chives and season with salt and white pepper. Cover bowl and place in refrigerator, or wrap in plastic film and freeze. Bring to room temperature before adding to hot grilled meat.

⸗ *Use with broiled steak, lamb, pork, or veal.*

MAÎTRE D'HÔTEL BUTTER

½ cup butter (softened)   salt and white pepper
2 tablespoons parsley (chopped)   1 tablespoon lemon juice

Cream the butter until it is fluffy. Add the parsley and season with salt and white pepper. Add the lemon juice gradually. When it is a smooth paste, wrap the mixture in plastic film and either refrigerate or freeze. Bring to room temperature before serving.

⸗ *This is a classic butter mixture that can be a delightful accompaniment to any grilled meat.*

SHALLOT BUTTER

8 shallots   salt and white pepper
½ cup butter (softened)

Place peeled shallots in just enough boiling water to cover. Parboil for about 5 minutes. Drain and cool, then mash the shallots until they are pulp. Cream the butter until fluffy and add the shallot pulp. Season with salt and white pepper. Cover bowl and place in refrigerator or wrap mixture in plastic film and freeze. Bring to room temperature before using.

⋆ *This delicately flavored (but actually quite hearty) butter is fine for any grilled meat, especially lamb chops.*

HERB BUTTER WITH VARIATIONS

Here is a recipe with which you can use your good imagination. It uses herbs that are in season—and butter, of course. If you have an herb garden, you can pluck a few herbs when they are young and tender. You can also buy certain herbs at vegetable markets and use them when they are fresh. You can use fresh dill, tarragon, basil, rosemary, thyme, even sage. Or you can devise a combination of 2 or more of these. The best way to preserve such seasonal herb butters is to freeze them. In this case, separate them into usable sizes and wrap (and mark their ingredients) with plastic film or any freezer type of paper. Of course, the following ingredients could be increased should you have enough fresh herbs on hand.

½ cup butter (softened)   salt and white pepper
4 tablespoons fresh herbs   dash of brandy
  (chopped)

Whip the butter with a fork until fluffy. Add the herbs and mix well. Season with salt and white pepper. Add the brandy and mix again.

⋆ *Use with any grilled meat.*

MUSTARD BUTTER

1 cup butter (softened)   dash of Worcestershire sauce
2 tablespoons dry mustard   ½ teaspoon sugar

Whip (or cream) the butter with a fork until it is fluffy. Stir in the mustard. When well blended, add the Worcestershire sauce and sugar. Cover and refrigerate, or wrap in plastic film and freeze. Bring to room temperature before serving, then whip again.

⸙ *This butter is a delight on grilled ham. It can also give a different taste to grilled steak or pork chops.*

CLARIFIED BUTTER

Even though we have not emphasized it in this book, clarified butter is the very best to use when you are sautéeing meat, but even excellent home cooks shy away from the idea. It seems troublesome, and they do not realize the advantages. The assets of using clarified butter are that there is no unattractive browning of butter while the meat is being sautéed, and there is little or no smoking. Actually, clarifying butter is quite simple.

Place ½ or 1 pound of butter in a heavy skillet over very low heat (actually, the top of a double boiler might be better). When butter is completely melted, you will find a sort of scum on top—a milky segment called whey. Scoop this off as well as you can and put it in a container (it might be useful later, as you can't help but get a bit of butter in the process). Now pour the clarified butter into another jar. Be extremely careful while pouring, as a slight residue has dropped to the bottom of the pan. Pour whatever residue is left into the container.

KNEADED BUTTER

This is a butter mixture that French chefs use quite a bit, and they call it *beurre manié*. They make it ahead of time and use it in an emergency when a quick thickening is needed and there is not time to make a roux. American cooks will find it equally satisfactory when the sauce or gravy is not quite thick enough. The best way to store kneaded butter is in sections that are about the size of a plum. Wrap them in plastic film and either refrigerate or freeze.

½ cup butter (softened)   1 cup flour

Beat the butter well, then add the flour gradually. When it has become a smooth paste, store in refrigerator or freezer. Do not season, as the sauce or gravy you will add it to will already have enough salt and pepper.

MUSHROOM SAUCE FOR MEAT

This is a sauce that is so delicious that it can be served all by itself as a vegetable. But it is really designed to be an excellent sauce for many types of meat.

¼ pound butter
1 medium onion (chopped)
2 stalks celery (chopped)
1 pound mushrooms (sliced)

½ cup Sauterne wine
salt and freshly ground pepper
¼ cup flour
¼ cup milk

Place butter in heavy skillet. When it is sizzling, add the onion and, a few minutes later, the celery. When they are slightly browned, add the mushrooms. Reduce heat and allow everything to simmer. When mushrooms are limp, add the wine and season with salt and pepper. Now combine the flour with milk and add gradually to the mushroom mixture, stirring well. When sauce is thickened, remove from heat. This sauce may be cooled and refrigerated for a week. Then it can be warmed up in a double boiler.

⸭ *Use with steak, calf's liver, or with chicken livers and rice. It would also be tasty served on crisp toast for a luncheon dish.*

VINAIGRETTE SAUCE

Here is a spicy sauce that requires no cooking and will store well in the refrigerator.

1 cup oil
3 tablespoons wine vinegar
1 tablespoon lemon juice
1 tablespoon parsley (minced)
1 tablespoon sour pickles (diced)
1 tablespoon pickle juice

1 tablespoon green pepper (diced)
2 tablespoons pimiento (chopped)
1 tablespoon capers (chopped)
2 cloves garlic (minced)
2 teaspoons sugar
salt and freshly ground pepper

Combine all of these ingredients in a jar, shake well, and refrigerate. Before serving, shake again. To arrive at a good combination of all of the ingredients, it is best to let them meld for two days.

⟍ *There are many uses for this sauce, but it is especially excellent when heated and served as a side dish for hot boiled meat.*

BASIC FRENCH DRESSING

Even though this is a salad dressing, it belongs in this cookbook, as it makes a fine marinade for various types of meats. The interesting thing about a basic dressing is that it can have a number of variations (just like béchamel sauce). For example, you can add a minced clove of garlic and have a garlic dressing. Or you can add a bit of dry mustard, celery seed, and oregano and have a zesty herb dressing. Just use your imagination.

4 tablespoons vinegar
12 tablespoons oil

salt and freshly ground pepper

See how easy? Just mix the vinegar with the oil and add salt and pepper to taste. Be sure to use your favorite vinegar, whether wine or plain cider, and your favorite oil. Some people insist on olive oil, others on sesame, corn, or whatever.

⸜ *Tip: A sprinkle of sugar might add something to this dressing. Another thing is that you can flavor your wine vinegar by inserting a sprig of fresh tarragon into the bottle. It looks so attractive.*

BARBECUE SAUCE AND MARINADE FOR POULTRY

½ cup oil
½ cup onions (finely chopped)
½ cup scallions (chopped)
¾ cup ketchup
2 cloves garlic (crushed)
¾ cup water
¼ cup lemon juice

3 tablespoons sugar
3 tablespoons Worcestershire sauce
2 tablespoons prepared mustard
salt and freshly ground pepper
2 tablespoons dry white wine

Heat oil in a heavy skillet and sauté the onions and scallions until lightly brown. Add the other ingredients and simmer for about 20 minutes, stirring from time to time. Remove from heat and allow to cool before using as a marinade.

⸜ *This sauce was especially designed as a marinade and then a basting sauce for chickens and ducks; however, it is also excellent for any veal you may wish to barbecue or broil in the oven.*

GIBLET SAUCE FOR POULTRY

This sauce is actually just the old-fashioned gravy accompaniment to roasted turkey, chicken, goose, or whatever. The old-time way was to use only the giblets from the bird to be roasted, but a more delightful sauce is achieved by buying extra giblets.

1 pound giblets
6 peppercorns
1 bay leaf
salted water

2 tablespoons butter
1 large onion (chopped)
2 tablespoons flour
pinch of sage

Place giblets (except livers and hearts) in sauce pan and add peppercorns and bay leaf, then add cold salted water just to cover. Boil over medium heat for 30 minutes. Add the livers and hearts and cook for another 15 minutes. Remove the giblets, strain the broth and reserve. When giblets are cool enough to handle, cut them in small pieces. Place butter in large skillet. When it is bubbling, add the onion and stir until golden brown. Move to one side of skillet and gradually stir in the flour with remaining butter. If butter has diminished, add a little more so that the flour makes a good mixture. Now slowly add the giblet broth, reducing heat. Add the giblets and add sage and taste to see if more salt is needed. When you have removed the roasted bird from the oven, add all of the pan drippings. If the sauce seems too thick, add a bit of beef bouillon.

MARROW SAUCE

4 marrow bones (1 inch long)	2 cloves garlic
1 cup Burgundy wine	2 teaspoons beef extract
4 sprigs parsley (chopped)	salt
1 medium onion (chopped)	1 teaspoon lemon juice
10 peppercorns	4 tablespoons butter (softened)
1 bay leaf	

Have butcher crack marrow bones. Soak bones in water to cover and leave in refrigerator overnight. Next day, remove marrow. Leave in large chunks and cover with cold water. Meanwhile, place wine, parsley, and onion in sauce pan. Tie the peppercorns, bay leaf, and garlic in cheesecloth bag and add to wine mixture. Place over medium heat, bring to boil, then add beef extract. Cook uncovered until mixture has diminished to ½ of volume (about 30 minutes). Remove cheesecloth bag, place rest of mixture in blender, and season with salt. Now place marrow pieces in boiling water with lemon juice, just enough to cover. Cook for about 3 minutes, strain, and add to wine sauce in blender. Add the softened butter and blend until sauce is smooth.

⁊ *The subtle flavor of this sauce makes it compatible with such delicate meats as filet mignon and prime ribs of beef.*

QUICK AND EASY STEAK SAUCE

steak drippings
3 tablespoons ketchup

2 tablespoons dry sherry
Worcestershire sauce

When steak has been cooked and removed to a hot platter, add ketchup and sherry to drippings in broiler pan. Then add Worcestershire sauce to taste (½ teaspoon or more), stir well, and place under broiler. Leave door open so that you can watch how sauce is cooking. It should be bubbling in 1 or 2 minutes. Remove from stove and pour over steak. This is a zesty sauce that's easy to make.

HOT MUSTARD AND OLIVE SAUCE FOR FONDUE

½ cup dry mustard
½ cup vinegar
12 seedless green olives
  (chopped)
1 tablespoon butter

1 egg (beaten)
¼ cup sugar
salt
¾ cup mayonnaise

Combine the mustard, vinegar, and olives in a bowl. Cover and let stand overnight in the refrigerator. Next day, allow this mixture to come to room temperature. Place butter in the top of a double boiler and allow it to melt. In a bowl, mix egg, sugar, and salt, beating well with a wire whisk. Add this mixture gradually to the warm butter and beat constantly. Gradually add the mustard and vinegar mixture and stir constantly. When the mixture actually coats a mixing spoon, remove from heat and allow to cool. About 20 minutes before serving, whip in the mayonnaise.

⁄ *This is such a hot and tangy sauce that it will add zip to any fondue you serve, but remember that it has a mayonnaise base and might be just dandy with cold cuts, with salad greens on the side.*

PICKLE AND CAPER SAUCE
FOR FONDUE

3 tablespoons dill pickles (minced)

2 tablespoons capers (drained and minced)

1 teaspoon prepared mustard

4 slices cooked bacon (pulverized)

1 tablespoon parsley (minced)

1 cup mayonnaise

1 tablespoon Basic French Dressing (pp. 290–291)

This is another sauce that needs no cooking. All of the ingredients are mixed well together and then placed with a cover in the refrigerator (3 hours or more) so that the flavors will blend well. Before serving time, this sauce should be brought to room temperature for at least half an hour and then stirred well.

⫟ *Aside from a fondue dip, this is great with every type of cold cut or as a sandwich spread.*

HORSERADISH AND BACON SAUCE
FOR FONDUE

This is a sauce that requires no cooking. It just needs several hours of refrigeration to bind it.

1½ cups sour cream

¼ cup white horseradish (drained)

¼ cup cooked bacon (crumbled)

1 tablespoon lemon juice

1 tablespoon sugar

1 teaspoon capers (chopped)

½ teaspoon dry mustard

1 tablespoon chives (minced)

Combine all ingredients in a bowl. Mix them well, cover the bowl, then place in refrigerator for at least 4 hours.

⫟ *Although this is an excellent dip for fondue, it can be delicious served with hot boiled tongue and broccoli.*

CURRY SAUCE FOR FONDUE

4 tablespoons butter	1 teaspoon flour
1 cup Spanish onions (minced)	¾ cup sweet heavy cream
2 cloves garlic (minced)	salt to taste
1½ teaspoons curry powder	

Melt the butter in a skillet. When bubbling, add the onions and garlic and stir until golden. Mix the curry powder (use more, if you are a curry fan) and the flour. Push the onions and garlic to one side and stir this into the butter. Gradually add the cream and stir everything together. Season with salt and allow to simmer (but do not boil) for about 5 minutes.

✦ *Even though this was devised as a fondue accompaniment, it can also be used to pep up leftover meat.*

GARLIC AND MUSHROOM SAUCE FOR FONDUE

3 tablespoons butter	1¼ cups beef bouillon
6 garlic cloves (crushed)	⅓ cup Madeira wine
2 tablespoons onions (minced)	salt
1 pound mushrooms (diced)	2½ tablespoons flour

Melt the butter in a heavy skillet and allow to bubble slightly. Add the crushed garlic, onions, and diced mushrooms. When mushrooms are limp, add 1 cup of the bouillon and increase heat so that mixture becomes somewhat reduced in volume. Add the wine and season with salt. Reduce heat and simmer. Now mix the flour with the ¼ cup of bouillon. Add this gradually to the mixture in skillet until it thickens. Be sure to stir so that the sauce is smooth.

✦ *Aside from a delectable fondue side dish, this sauce can be used for any type of broiled or sautéed meat.*

ROQUEFORT AND ALMOND SAUCE
FOR FONDUE

¾ pound Roquefort cheese
¼ cup almonds (finely
  chopped)
¾ cup butter (soft)

½ teaspoon prepared mustard
few drops angostura bitters
½ cup heavy sweet cream

Here is a recipe that requires no cooking, just mixing
and refrigerating. The Roquefort cheese should be brought to
room temperature until it is quite soft. Then it should be combined with the rest of the ingredients and mixed well. The
mixture is placed covered in the refrigerator for 3 hours or
more. Before serving, this sauce should be brought to room
temperature for 2 hours so that it will be soft enough to whip
to a fluff.

Index

Alabama Pig's Feet, 174
All-American Veal Loaf, 113
Alston, Elizabeth, 82
Arkansas Meat Loaf, 72
Arm lamb chops, 123
Arm roast of ham, 151
Auchincloss, Mr. and Mrs. Douglas, 46
Avedon, Richard, 132

Baby Lamb, 116
    Broiled, with Soy Sauce and Herbs,
        128–129
Bacon
    Broiled Brains and, 246–247
    Kentucky Hush Puppies with, 173
    Puffs, Liver and, 236–237
    slabs, 156
Baked Chicken Halves with Wine, 190–
    191
Baked Corned Beef Hash with Cabbage,
    89
Baked Fillet of Beef, 51
Baked Fresh Ham with White Wine,
    158–159
Baked Ham Steak with Pineapple, 161
Baked Ham Steak with Sherry, 162
Baked Ham with Beer, 160
Baked Ham with Pink Champagne, 159–
    160
Baked Virginia Smithfield Ham, 157
Barbecue Sauce and Marinade for Poul-
    try, 291
Barbecued Chicken No. 1, 184–185
Barbecued Chicken No. 2, 185
Barbecued Lamb Chops, 136–137
Basic French Dressing, 290–291
Basic Fresh Beef Tongue, 240
Béarnaise Sauce, 284
Beaujolais Beef Stew, 83

Béchamel Sauce, 282
Beef, 29–89
    aging of, 30
    Baked Fillet of, 51
    -Ball Canapés, Hot, 67
    buying, 30–45
        brisket, 44–45
        chuck, ground, 40
        chuck roast, 44
        chuck steak, 38–39
        club (Delmonico) steak, 35–36
        color and texture, 30–31
        cuts (how to identify), 33–45
        fillet strips, 36
        flank steak, 39
        forequarter, 33
        ground, 40–41
        hindquarter, 34
        kosher, 45
        porterhouse steak, 34
        prime quality, 30–31
        prime ribs, 41–42
        rib steak, 38
        roasts, 41–45
        round, ground, 40
        round steak, 39–40
        rump roast, 43–44
        shell roast, 42
        shell steak, 36–37
        short ribs, 45
        shoulder roast, 43
        sirloin, ground, 40
        sirloin roast, 43
        sirloin steak, 37
        sirloin tip steak, 38
        steaks, 34–40
        tail of porterhouse, ground, 40–41
        T-bone steak, 35
        tenderloin steak, 36
        what to avoid, 31

Beef (cont.)
  ground, 40–41
  kidneys, 233
  liver, 232
  prime
    marbling (or graining), 31
    yield per steer, 30
  recipes, 46–89
  roasts, 41–45
  steaks, 34–40
  stew, 78–79, 81–84
  sweetbreads and brains, 234–235
  Tongue, 234
    Basic Fresh, 240
    Fresh, with Vegetable Sauce, 241
    and Ham with Orange, 240–241
  types of cattle, 29–30
Beef and Veal Goulash, 110
Beef and Vegetable Soup, 73
Beef Blintzes, 65
Beef Burgundy, 85
Beef Fondue Bourguignonne, 61–63
Beef Orientale, 55
Beef Scallopini Roulades, 52–53
  Stuffing for, 281
Beef Scallopini with Prosciutto, 53
Beef Stew with Beer, 82
Beef Stew with Wine, 78–79
Beef Stroganoff, 58
Beef Wellington, 50
Benenson, Edward H., 221
Bennett, Joan, 125
Berns, Charlie, 50
Berns, H. Jerome, 50
Beurre Noir Sauce, 285
Birds, defrosting, 11
Blade lamb chops, 123
Blade loin roast pork, 155
Blade pork chops, 155
Blade pork steaks, 152
Blintzes, Beef, 65
Bloom, Claire, 112
Boneless Capon with Chestnut Stuffing, 211–212
Boneless smoked shoulder butt, 152
Boning knife, 13
Bordelaise Sauce No. 1, 284
Bordelaise Sauce No. 2, 285
Boston butt, 152
Brain Salad, 246
Brains
  Broiled, and Bacon, 246–247
  Calf's, with Capers, 247

Brains (cont.)
  Sautéed, and Mushrooms, 248
  and sweetbreads
    beef, 235
    calf's, 234–235
Braised Chicken with Raisins, 188
Braised Chinese Duck, 222–223
Braised Duck with Giblet Sauce, 224–225
Braised Duck with Wine, 223
Braised Flanken (Short Ribs of Beef), 74
Braised Lamb Shanks, 138
Braised Lamb Shanks with Tomato Sauce, 138–139
Braised Rock Cornish Game Hen, 228–229
Braised Swiss Steak, 60–61
Braised Venison, 266–267
Bread Stuffing for Poultry, 272
Breaded Veal Scallopini No. 1, 100
Breaded Veal Scallopini No. 2, 101
Breast of veal, 99
Brisket of beef, 44–45
Brisket of Beef, 80
Broiled Baby Lamb with Soy Sauce and Herbs, 128–129
Broiled Brains and Bacon, 246–247
Broiled Chicken Halves, 185–186
Broiled Double Lamb Chops, 130
Broiled Eggplant, Ham, Tomato, and Cheese Combo, 174–175
Broiled Fillet Strips with Fantastic Sauce, 46–47
Broiled Lamb Steaks, 136
Broiled Quail, 259
Broiled Venison Chops, 268
Burger-Furter, 70–71
Butter
  Chive, 286
  Clarified, 288
  Herb, with Variations, 287
  Kneaded, 289
  Maître d'Hôtel, 286
  mixtures, 286
  Mustard, 288
  Shallot, 287
Butterfly Filet Mignon with Sherry, 47
Butterfly Leg of Lamb, 127

Calf's Brains with Capers, 247
Calf's Liver, 232
  Sautéed
    with Herbs, 235–236
    with Onion Sauce, 236

Calf's sweetbreads and brains, 234–235
Calf's tongue, 234
Calves, types of, 90
Canned meats, care of, 7
Capon
  Boneless, with Chestnut Stuffing, 211–212
  facts about, 180
  Roasted
    No. 1, 210
    No. 2, 210–211
Carver's helper, 15
Carving meat and poultry, 12–28
  care of cutlery, 16
  hams, 16, 22–24
    butt half, 24
    shank half, 23
    whole, 22
  implements, 13–15
    boning knife, 13
    carver's helper, 15
    poultry shears, 14
    roast carver, 15
    standard set, 14
    steak set, 14–15
  and knife-sharpening, 15–16
  large fowl, 26–28
  platter, 15
  roasts, 16–22
    beef, 16–19
    beef Delmonico (rib eye), 18
    blade bone pot, of beef, 19
    crown, 22
    lamb, 20
    leg of lamb, 20
    loin, of pork, 21
    pork, 21
    rolled, 18–19
    rolled rib, of beef, 18
    standing rib, of beef, 16–17
  steaks, 25
    porterhouse, 25
    T-bone, 25
  tips on, 13
  turkey, 26–28
Casserole(s)
  Chicken
    Flambé, 192
    Jubilee (Flambé), 191
  Duck, 225
  Frankfurter and Baked Bean, 176
  Lamb
    Succulent, with Vegetables, 144–145

Casserole(s)
  Lamb (cont.)
    Kidney and Mushroom, 242
  Pigeon (or Squab), 265
  Pork, with Cognac, 168
  Rabbit
    No. 1, 268–269
    No. 2, 269
    No. 3, 270
  Tripe, 249
  Veal
    and Green Pepper, Simple, 112–113
    Hearts and Red Wine, 248–249
    and Sour Cream, 110–111
    and Sweetbread, 245
    and Vegetable, 109
Cattle, types of, 29–30
Center loin pork roast, 153
Center section of hog, 153–156
Center veal leg roast, 93
Chaîne des Rotisseurs (dining society), 55, 221
Chestnut and Brazil Nut Stuffing for Poultry, 275
Chestnut Stuffing for Poultry, 274–275
Chicken, 179–180
  Alabama Style with Cream Sauce, 195
  Almond Soup, 203
  Balls
    Paprika, with Sherry, 205
    with Pumpkin Puree, 204–205
  Barbecue, Meat and (Oven-Cooked), 169–170
  Barbecued
    No. 1, 184–185
    No. 2, 185
    Sweet and Sour, 185
  Braised, with Raisins, 188
  Breasts
    Crispy Oven-Fried, 189–190
    with Orange Sauce, 187
  Cacciatore, 202–203
  Casserole
    Flambé, 192
    Jubilee (Flambé), 191
  Coq Au Vin—the Easy Way, 199
  Creamed
    —Southern Style, 194
    and Sweetbreads, 209
  Croquettes, 206–207
  Curried, 200
  Cutlets, Easy, 206
  Deep-Fried, in Batter, 190

Chicken (cont.)
  defrosting, 10
  facts about, 179–181
    capon, 180
    Rock Cornish game hens, 180–181
  Fricassee, 198
  Gumbo—with a New Orleans Flavor, 202
  Halves
    Baked with Wine, 190–191
    Broiled, 185–186
  Jubilee Casserole (Flambé), 191
  mature, 179
  Meat and, Barbecue (Oven-Cooked), 169–170
  Oven-Fried
    Crispy Breasts, 189–190
    No. 1, 188
    No. 2, 189
  Salad
    No. 1, 207
    No. 2, 208
    No. 3, 208–209
  Sesame, 196
  Stewed, with Dumplings, 197
  and Tomato Soup with Wine, 204
  and Tomato Stew with Curry, 201
  in White Wine, 198–199
  and Wine Casserole with Apples, 192–193
  young tender-meated, 179
Chicken Alabama Style with Cream Sauce, 195
Chicken Almond Soup, 203
Chicken and Tomato Soup with Wine, 204
Chicken and Tomato Stew with Curry, 201
Chicken and Wine Casserole with Apples, 192–193
Chicken Balls with Pumpkin Puree, 204–205
Chicken Breasts with Orange Sauce, 187
Chicken Cacciatore, 202–203
Chicken Casserole Flambé, 192
Chicken Croquettes, 206–207
Chicken Fricassee, 198
Chicken Gumbo—with a New Orleans Flavor, 202
Chicken in White Wine, 198–199
Chicken Jubilee Casserole (Flambé), 191
Chicken Legs with Soy Sauce, 193
Chicken Liver Dumplings, 238

Chicken Livers with Wine, 237
Chicken Salad No. 1, 207
Chicken Salad No. 2, 208
Chicken Salad No. 3, 208–209
Chicken Sesame, 196
Chile con Carne, West Indian, 68–69
Chinese Duck, Braised, 222–223
Chitterlings, 250–251
Chive Butter, 286
Chopped Steak Layered with Onion and Pepper Rings, 71
Chuck
  ground, 40
  roast, 44
  steak, 38–39
Clarified Butter, 288
Classic Broiled Porterhouse Steak, 46
Club (Delmonico) steak, 35–36
Collard Greens and Ham Hocks, 172–173
Columbus, Christopher, 146
Coq au Vin—the Easy Way, 199
Corned Beef Hash with Cabbage, Baked, 89
Crazy Beef Stew, 84
Creamed Chicken—Southern Style, 194
Creamed Chicken and Sweetbreads, 209
Creamed Sweetbreads Maryland No. 1, 244
Creamed Sweetbreads Maryland No. 2, 244–245
Creamy Hamburgers—Sautéed, 70
Creamy Mustard Sauce for Ham, 283
Crispy Oven-Fried Chicken Breasts, 189–190
Crown Loin of Pork
  Fruit Stuffing for, 279–280
  Meat and Sausage Stuffing for, 280
  Roasted, 164–165
Crown of Lamb
  Meat Stuffing for, 279
  Stuffed, 133
Crown roast of lamb, 122
Crown roast of veal, 97
Crown roasts, carving of, 22
Cured meats, care of, 7
Curried Chicken, 200
Curried Lamb, 140
Curried Lamb Chops, 135
Curry Sauce for Fondue, 295
Cutlery
  care of, 16
  for carving, 13–16
  sharpening knives, 15–16

Date and Nut Stuffing for Poultry, 276
Deep-Fried Chicken in Batter, 190
Defrosting, *see* Storage and refrigeration, defrosting
Delmonico (club) steak, 35–36
De Soto, Hernando, 146
Dilled Veal, 111–112
Double lamb roast, rolled, 120
Double Rib Rack of Lamb (Twin Lamb Eyes), 131
Double sirloin roast, rolled, 95
Doves, 254
Duck, 183, 253–254
  Braised
    Chinese, 222–223
    with Giblet Sauce, 224–225
    with Wine, 223
  buying tips, 183
  Casserole, 225
  facts about, 183
  mature, 179
  Roast
    with Apples, 220–221
    Flambé No. 1, 218–219
    Flambé No. 2, 219
    with Fruit and Liqueur Sauce, 221–222
    with Orange, 220
  Salad, with Fruit and Nuts, 226
  Stewed, with Chestnuts, 226
  Wild, 253–254
    Roast, No. 1, 258
    Roast, No. 2, 258–259
  young tender-meated, 179
Duck Casserole, 225
Duck Salad with Fruit and Nuts, 226
Dumplings, Chicken Liver, 238

Easy Chicken Cutlets, 206
Easy London Broil, 51-52
Eggplant, Ham, Tomato, and Cheese Combo, Broiled, 174–175
Eggs Benedict, 175
English lamb chops, 121

Fagiano alla Crema (Pheasant in Cream), 263
Family Pot Roast, 77–78
Felder, Raoul Lionel, 206

Filet Mignon (tenderloin steak), 36
  Beef Wellington, 50
  Butterfly, with Sherry, 47
  with Cognac and Herb Sauce, 49–50
  with Peppercorns and Brandy, 48–49
  Tournedos of Beef with Béarnaise Sauce, 48
Filet Mignon with Cognac and Herb Sauce, 49–50
Filet Mignon with Peppercorns and Brandy, 48–49
Fillet of Beef, Baked, 51
Fillet Strips, 36
  Broiled, with Fantastic Sauce, 46–47
Fillet Tidbits with Yams, 57
Flank Steak, 39
  Grilled, 52
Flanken (short ribs), 45
  Braised, 74
Flanken Stew, 83–84
Fondue, Sauce for, 293–296
  Curry, 295
  Garlic and Mushroom, 295
  Horseradish and Bacon, 294
  Hot Mustard and Olive, 293
  Pickle and Caper, 294
  Roquefort and Almond, 296
Foreshank of veal, 98
Fowl
  carving, 26–28
  defrosting
    chicken sections, 10
    large turkeys, 10–11
    small birds, 11
  kosher quality, 45
  mature, 179
  storage and refrigeration, 6, 7
  young tender-meated, 179
    chicken, 179
    duck, 179
    turkey, 179
Frankfurter and Baked Bean Casserole, 176
Franklin, Benjamin, 181–182
French Dressing, Basic, 290–291
French rib lamb chops, 122
Frenched rib veal chops, 98
Fresh Beef Tongue with Vegetable Sauce, 241
Fresh Ham with White Wine, Baked, 158–159
Fresh meat, care of, 6–7
Fried Tripe, 250

Frozen meat and fowl, care of, 7
Fruit, Nut, and Sausage Stuffing for
    Poultry, 276–277
Fruit Stuffing for Crown Loin of Pork,
    279–280

Game, see Wild birds and animals
Garlic and Mushroom Sauce for Fon-
    due, 295
Giblet Fricassee, 239
Giblet Sauce for Poultry, 291–292
Giblet Stuffing for Poultry, 277
Gigot (Lamb Roasted French Style), 126
Goose, 184, 253
    buying tips, 184
    facts about, 184
    Roast, with Burgundy, 230
    Wild, 253
        Roasted, 257
Graham, Virginia, 196
Greek Grape Leaves Stuffed with Lamb,
    143
Grilled Flank Steak, 52
Ground beef, 40–41
    chuck, 40
    round, 40
    sirloin, 40
    tail of porterhouse, 40–41
Grouse, 254
    Roasted, 262
Guinea Hen
    facts about, 181
    with Wild Rice Stuffing, Roasted, 229
Gussow, Don, 85

Halft standing rib roast No. 1, 41
Half standing rib roast No. 2, 42
Ham
    Baked
        with Beer, 160
        with Pink Champagne, 159–160
        Virginia Smithfield, 157
    butt, 150
    carving, 16, 22–24
        butt half, 24
        shank half, 23
        whole, 22
    Creamy Mustard Sauce for, 283
    for picnic, 151
    shank, 150

Ham (cont.)
    and Split Pea Soup, 171
    Spring Leg of Lamb—Roasted with,
        128
    Steak, 150
        Baked, with Pineapple, 161
        Baked, with Sherry, 162
        with Orange, 162
        with Peanut Butter and Grapes,
            161
    Tongue and, with Orange, 240–241
    whole, 149
    See also Pork
Ham and Split Pea Soup, 171
Ham Bone and Split Pea Soup, 172
Ham Hocks and Collard Greens, 172–
    173
Ham Steak with Orange, 162
Ham Steak with Peanut Butter and
    Grapes, 161
Hamburger(s)
    Burger-Furter, 70–71
    Creamy—Sautéed, 70
    Juicy—Broiled, 69
Hash, Turkey, 217
Heart, 235
Herb Butter with Variations, 287
Hillman, Libby, 266
Hindsaddle of veal, 93–95
Hog cuts (how to identify), 149–156
Hogs, 148
Hollandaise Sauce, 283
Home freezing, 8–9, 10
    "keeping power," 9
    storage time for, 9–10
    thawed meat and, 10
    See also Storage and refrigeration
Horseradish and Bacon Sauce for Fon-
    due, 294
Hot Beef-Ball Canapés, 67
Hot Mustard and Olive Sauce for Fon-
    due, 293
Hothouse Lamb, 116
    Roasted, 134–135
Huste, Anne-Marie, x

Involtini di Vitello (Stuffed Scallopini),
    102

Juicy Hamburgers—Broiled, 69

Kennedy, Jacqueline, x
Kentucky Hush Puppies with Bacon, 173
Kidney, 232–233
  beef, 233
  lamb, 233
  pork, 233
  veal, 233
    chops, 96
Kneaded Butter, 289
Knickerbocker, Suzy, x
Knife-sharpening, 15–16
Korean Bulgogi (Sirloin Strips), 54–55
Kosher products, 45
Kriendler, I. Robert, 50
Kriendler, Jack, 50
Kwit, Dr. Nathaniel Troy, 55

Lamb, 115–145
  Baby, 116
    Broiled, with Soy Sauce and Herbs, 128–129
  and Bacon Meatballs, 142
  basic cuts of, 117
  buying, 117–124
    arm chops, 123
    blade chops, 123
    crown roast, 122
    cuts (how to identify), 118–124
    English chops, 121
    French rib chops, 122
    kebabs, 119
    kidneys, 233
    leg, 118–119
    leg chops (or steaks), 119
    liver, 232
    loin, 120–121
    loin chops, 120–121
    neck, 124
    rack, 121–122
    rack or rib roast, 121
    rib chops, 122
    riblets, 124
    roast loin, 120
    roasts, 118–119, 120, 121–122
    rolled double roast, 120
    rolled leg, 119
    shank and breast, 123–124
    shank half of leg, 118
    shoulder, 123–124
    sirloin half of leg, 119

Lamb
  buying (*cont.*)
    tips on, 117
    tongue, 234
    what to avoid, 117
    whole leg, 118
  Casserole with Vegetables, Succulent, 144–145
  Chops
    arm, 123
    Barbecued, 136–137
    blade, 123
    Curried, 135
    Double Broiled, 136
    English cut, 121
    French rib, 122
    leg (or steaks), 119
    loin, 120–121
    rib, 122
  Curried, 140
    Chops, 135
    Roast, Leg of, 129–130
  Gigot (Roasted French Style), 126
  Greek Grape Leaves Stuffed with, 143
  Hothouse, 116
    Roasted, 134–135
  and Kidney Bean Stew, 140–141
  Leg of
    Butterfly, 127
    Roast, with Special Bennett Paste, 125
    Roast Curried, 129–130
  Noisettes of, 137
  in Pepper Cups, 144
  Rack of
    Double Rib (Twin Lamb Eyes), 131
    Roast, with Marmalade, 131
    Roasted Stuffed, 132–133
  recipes, 124–145
  regular, 116
  Saddle of, Roasted with Peach Nectar, 124–125
  Shanks
    Braised, 138
    Braised, with Tomato Sauce, 138–139
  Shish Kebab (or Shashlik), 139
  Shoulder of, Roast with Apricots, 130
  Spring Leg of
    —Roasted with Currant Jelly, 126–127
    —Roasted with Ham, 128

Lamb (cont.)
  Steaks, Broiled, 136
  Stew, Savory, 141
  Stuffed Crown of, 133
  types of sheep, 115–116
  yearling, 116
    mutton, 116–117
Lamb and Bacon Meatballs, 142
Lamb and Kidney Bean Stew, 140–141
Lamb in Pepper Cups, 144
Lamb Kidney and Mushroom Casserole, 242
Lamb Shish Kebab (or Shashlik), 139
Lamburgers with Mushroom Caps, 142–143
Large fowl, carving, 26–28
Large turkeys, defrosting, 10–11
Leftover meats, care of, 8
Leg lamb chops (or steaks), 119
Leg of lamb, 118–119
Leg of pork, 149–150
Liver, 231
  and Bacon Puffs, 236-237
  beef, 232
  calf's, 232
  chicken, 232
  lamb, 232
  pork, 232
  Sautéed
    with Herbs, 235–236
    with Onion Sauce, 236
  varieties of, 231–232
Liver and Bacon Puffs, 236–237
Lobel, Leon, x–xiii
Lobel, Morris, xi–xii
Lobel, Nathan, xii
Lobel, Stanley, x–xiii
Loin lamb chops, 120–121
Loin of lamb, 120–121
Loin of Pork
  North Carolina Roast, 163
  Roast, 163
    —Sweet and Pungent, 164
  Roasted Crown, 164–165
Loin of Veal, 95–96
  Fine Herbs, Roasted, 107
  rolled roast, 96
Loin pork chops, 154
Loin roast of veal, 95
Loin veal chops, 96
London Broil, Easy, 51–52
Long Island Duck Farmers Cooperative, Inc., 183

Maître d'Hôtel Butter, 286
Makowsky, Jacques, 180–181
Marrow Sauce, 292
Meade, Julia, 107
Meat(s)
  beef, 29–89
    recipes, 46–89
  carving, see Carving meat and poultry
  consumption per person (U.S.), 2
  defrosting, see Storage and refrigeration, defrosting
  facts about buying, 1–5
  government inspection of, 2–3
  grades of, 3–4
  kosher, 45
  lamb, 115–145
    recipes, 124–145
  pork, 146–176
    recipes, 157–176
  sauces, 282–296
  storage and refrigeration, see Storage and refrigeration
  stuffings, 271–281
  tenderizing, 11
  variety, 231–251
    recipes, 235–251
  veal, 90–114
    recipes, 99–114
  wild birds and animals, 252–270
    recipes, 256–270
Meat and Chicken Barbecue (Oven-Cooked), 169–170
Meat and Sausage Stuffing for Crown of Pork, 280
Meat delicacies, 5
Meat Loaf, Arkansas, 72
Meat Loaf Parmigiana, 73
Meat Stuffing for Crown of Lamb, 279
Meatballs
  —Baked, Stuffed, 66–67
  Lamb and Bacon, 142
  Swedish, 66
Mushroom Sauce for Meat, 289
Mustard Butter, 288

Namath, Joe, 129
Neck of veal, 98
Noisettes of Lamb, 137
North Carolina Roast Loin of Pork, 163

O'Brien, Robert H., 78

Old Dutch Wiener Schnitzel, 99
Old-fashioned Beef Stew, 81
Onassis, Aristotle, x
Orthodox Jews, 45
Osso Buco, 85–86
Oven-Fried Chicken Breasts, Crispy, 189–190
Oven-Fried Chicken No. 1, 188
Oven-Fried Chicken No. 2, 189
Oxtail Ragout No. 1, 86–87
Oxtail Ragout No. 2, 87
Oxtail Stew, 88–89
Oyster Stuffing — Baltimore Style — for Poultry, 278

Paprika Chicken Balls with Sherry, 205
Partridge, 254
   with Cabbage (Perdrix aux Choux), 260
Partridge—Spanish Style, 261
Pheasant, Roast, 262–263
Pheasant in Cream (Fagiano alla Crema), 263
Pickle and Caper Sauce for Fondue, 294
Picnic ham, 151
Pigeon, 254
Pigeon (or Squab) Casserole, 265
Pigs, 148
Pig's Feet, 151
   Alabama, 174
Pork, 146–176
   Baked Fresh Ham with White Wine, 158–159
   buying, 147–156
      arm roast of ham, 151
      bacon slabs, 156
      blade chops, 155
      blade loin roast, 155
      blade steaks, 152
      boneless smoked shoulder butt, 152
      Boston butt, 152
      center loin roast, 153
      center section of hog, 153–156
      cuts (how to identify), 149–156
      foreleg (or shoulder), 150–152
      ham butt, 150
      ham shank, 150
      ham steak, 150
      hocks, 151
      kidneys, 233
      leg, 149–150
      liver, 232

Pork
   buying (cont.)
      loin chops, 154
      picnic ham, 151
      pig's feet, 151
      rib chops, 154
      rolled loin roast, 154
      salt, 156
      sirloin chops, 153
      sirloin roast, 153
      spareribs, 155–156
      tips on, 147–148
      tongue, 234
      what to avoid, 148
   Casserole with Cognac, 168
   Chops
      with Apple Brandy, 167
      blade, 155
      loin, 154
      in Orange Marinade, 166
      with Pineapple, 165
      in Red Wine, 166
      rib, 154
      sirloin, 153
   quality of, 147
   recipes, 157–176
   Roasted Suckling Pig, 158
   swine production, 147
   types of swine, 148
   See also Ham
Pork and Sauerkraut Stew, 167–168
Pork Casserole with Cognac, 168
Pork Chops in Orange Marinade, 166
Pork Chops in Red Wine, 166
Pork Chops with Apple Brandy, 167
Pork Chops with Pineapple, 165
Porterhouse Steak, 34
   carving, 25
   Classic Broiled, 46
Pot Roast
   Family, 77-78
   Tomato, 79
Potato Stuffing for Poultry, 278
Potofsky, Jacob S., 83
Poultry, 177–230
   Barbecue Sauce and Marinade for, 291
   buying, 177–184
      chicken, 179–180
      duck, 183
      goose, 184
      government grading and, 177–178
      guinea hen, 181

Poultry
  buying (cont.)
    kosher, 45
    mature fowl, 179
    turkey, 181–182
    young tender-meated fowl, 179
  carving, see Carving meat and poultry
  defrosting, see Storage and refrigeration, defrosting
  Giblet Sauce for, 291–292
  home freezing, 8–9
    storage time, 9
  recipes, 184–230
  sauces, 282–296
  shears, 14
  storage and refrigeration, see Storage and refrigeration
  stuffings, 271–281
  See also Wild birds and animals
Prime beef
  marbling (or graining), 31
  yield per steer, 30
Prime ribs of beef, 41–42
  half standing roast
    No. 1, 41
    No. 2, 42
  rolled rib roast, 42
  standing rib roast, 41
Pynchon, William, 146

Quail, 254
  Broiled, 259
Quick and Easy Steak Sauce, 293
Quick Creamed Veal, 114

Rabbit, 255–256
Rabbit Casserole No. 1, 268–269
Rabbit Casserole No. 2, 269
Rabbit Casserole No. 3, 270
Rack of Lamb, 121–122
  Double Rib (Twin Lamb Eyes), 131
  Roast, with Marmalade, 131
  Roasted Stuffed, 132–133
Rack or rib roast of lamb, 121
Ragout of Beef, 80–81
Refrigeration, see Storage and refrigeration
Regular lamb, 116
Reingold, Carmel Berman, 88
Reingold, Harry, 88
Rib lamb chops, 122
Rib pork chops, 154

Rib roast
  of lamb, 121
  rolled, 42
  standing, 41
  of veal, 97
Rib Roast, Standing, 74
Rib of veal, 96–98
Rib steak, 38
Rib veal chops, 97
  Frenched, 98
Riblets, lamb, 124
Rivers, Joan, 105
Roaman, Carol, 104
Roaman, Martin, 104
Roast(s)
  beef
    brisket, 44–45
    carving, 16–19
    chuck, 44
    prime ribs, 41–42
    rump, 43–44
    shell, 42
    shoulder, 43
    sirloin, 43
  carver, 15
  carving, see Carving meat and poultry, roasts
  defrosting, 10
  lamb
    carving, 20
    crown, 122
    leg, 118–119
    loin, 120
    rack, 121–122
    rack or rib, 121
    rolled double, 120
    rolled leg, 119
    shank half of leg, 118
    sirloin half of leg, 119
    whole leg, 118
  veal
    center leg, 93
    crown, 97
    hindsaddle, 93–95
    loin, 95–96
    rib, 96–98
    rolled double sirloin, 95
    rolled leg, 93
    rolled loin, 96
    shank half of leg, 93–94
    sirloin, 94–95
    standing sirloin, 95
  See also Ham; Pork; Poultry; Stuffing(s); specific recipes

Roast Curried Leg of Lamb, 129–130
Roast Duck Flambé No. 1, 218–219
Roast Duck Flambé No. 2, 219
Roast Duck with Apples, 220–221
Roast Duck with Fruit and Liqueur
   Sauce, 221–222
Roast Duck with Orange, 220
Roast Goose with Burgundy, 230
Roast Leg of Lamb with Special Ben-
   nett Paste, 125
Roast loin of lamb, 120
Roast Loin of Pork, 163
   North Carolina, 163
Roast Loin of Pork—Sweet and Pungent,
   164
Roast Pheasant, 262–263
Roast Rack of Lamb with Marmalade,
   131
Roast Shoulder of Lamb with Apricots,
   130
Roast Squab No. 1, 264
Roast Squab No. 2, 264–265
Roast Turkey Breast, 216
Roast Turkey Hen No. 1, 212–213
Roast Turkey Hen No. 2, 213
Roast Wild Duck No. 1, 258
Roast Wild Duck No. 2, 258–259
Roast Wild Turkey, 256–257
Roasted Capon No. 1, 210
Roasted Capon No. 2, 210–211
Roasted Crown Loin of Pork, 164–165
Roasted Grouse, 262
Roasted Guinea Hen with Wild Rice
   Stuffing, 229
Roasted Hothouse Lamb, 134–135
Roasted Loin of Veal Fine Herbs, 107
Roasted Saddle of Lamb with Peach
   Nectar, 124–125
Roasted Stuffed Rack of Lamb, 132–133
Roasted Suckling Pig, 158
Roasted Wild Goose, 257
Rock Cornish Game Hen(s)
   Braised, 228–229
   Breasts (Kiev), 228
   facts about, 180–181
   with Herb and Shallot Butter, 227
   —Roasted, 227
Rock Cornish Game Hen—Roasted, 227
Rock Cornish Game Hen Breasts (Kiev),
   228
Rock Cornish Game Hen with Herb and
   Shallot Butter, 227
Rockefeller, Mr. and Mrs. Winthrop, 72
Rolled double lamb roast, 120

Rolled double sirloin veal roast, 95
Rolled leg of lamb, 119
Rolled leg of veal, 93
Rolled loin pork roast, 154
Rolled loin veal roast, 96
Rolled rib roast, 42
Rolled roasts, carving of, 18–19
Roquefort and Almond Sauce for Fon-
   due, 296
Round, ground, 40
Round Roast of Beef with Vegetables, 75
Round steak, 39–40
Round veal steak (or cutlet), 94
Rump roast, 43–44
Rump Roast of Beef, 76–77

Sage and Onion Stuffing for Poultry,
   271–272
Salt pork, 156
Saltimbocca (Veal Rolls), 106
Sandifer, Jawn A., 68–69, 163
Sandifer, Mrs. Jawn A., 68–69
Sauce(s), 282–296
   Barbecue, and Marinade for Poultry,
      291
   Basic French Dressing, 290–291
   Béarnaise, 284
   Béchamel, 282
   Beurre Noir, 285
   Bordelaise
      No. 1, 284
      No. 2, 285
   Butter, 286–289
      Chive, 286
      Clarified, 288
      Herb, with Variations, 287
      Kneaded, 289
      Maître d'Hôtel, 286
      mixtures, 286
      Mustard, 288
      Shallot, 287
   Creamy Mustard, for Ham, 283
   for Fondue, 293–296
      Curry, 295
      Garlic and Mushroom, 295
      Horseradish and Bacon, 294
      Hot Mustard and Olive, 293
      Pickle and Caper, 294
      Roquefort and Almond, 296
   Giblet, for Poultry, 291–292
   Hollandaise, 283
   Marrow, 292
   Mushroom, for Meat, 289

Sauce(s) (cont.)
 Steak, Quick and Easy, 293
 Vinaigrette, 290
Sauerbraten, 76
Sauerkraut Stew, Pork and, 167–168
Sautéed Brains and Mushrooms, 248
Sautéed Calf's Liver with Herbs, 235–236
Sautéed Calf's Liver with Onion Sauce, 236
Savory Kasha Stuffing for Poultry, 274
Savory Lamb Stew, 141
Savory Wild Rice Stuffing for Poultry, 273
Scallopini, 94
 Beef
  with Prosciutto, 53
  Roulades, 52–53, 281
 Veal
  Breaded, No. 1, 100
  Breaded, No. 2, 101
  Involtini di Vitello (Stuffed), 102
  Marsala, 101–102
  Mozzarella, 103
Scallopini of Turkey Breast Marsala, 214
Scallopini of Veal Marsala, 101–102
Shallot Butter, 287
Shank half of leg of lamb, 118
Shank half of leg of veal, 93–94
Sheep, types of, 115–116
Shell roast, 42
Shell Roast New Yorker, 75
Shell steak, 36–37
Shish Kebab (or Shashlik), 139
Short ribs (flanken), 45
Short Ribs of Beef (Flanken), Braised, 74
Shoulder butt, boneless smoked, 152
Shoulder of veal, 98
Shoulder roast, 43
Simple Veal and Green Pepper Casserole, 112–113
Sirloin
 ground, 40
 half of leg of lamb, 119
 pork chops, 153
 roast, 43
  of pork, 153
  of veal, standing, 95
 steak, 37
  tip, 38
 of veal, 94–95
Sirloin Strips (Korean Bulgogi), 54–55

Small birds, defrosting, 10–11
Smoked meats, care of, 7
Spareribs, 155–156
 Texan, 170–171
Spareribs with Wine Vinegar, 170
Split Pea Soup
 Ham and, 171
 Ham Bone and, 172
Spring Leg of Lamb—Roasted with Currant Jelly, 126–127
Spring Leg of Lamb—Roasted with Ham, 128
Squab, 254
 Casserole, 265
 or Pigeon Casserole, 265
 Roast
  No. 1, 264
  No. 2, 264–265
Standard carving set, 14
Standing Rib Roast, 74
Standing rib roast (cut), 41
Standing sirloin roast of veal, 95
Steak(s)
 in the Blanket, 61
 buying, 34–40
  chuck, 38–39
  club (Delmonico), 35–36
  fillet strips, 36
  flank, 39
  porterhouse, 34
  rib, 38
  round, 39–40
  shell, 36–37
  sirloin, 37
  sirloin tip, 38
  T-bone, 35
  tenderloin, 36
 carving, 25
  porterhouse, 25
  set, 14–15
  T-bone, 25
 Chopped, Layered with Onion and Pepper Rings, 71
 cuts (how to identify), 34–39
 defrosting, 10
 Filet Mignon
  Beef Wellington, 50
  Butterfly, with Sherry, 47
  with Cognac and Herb Sauce, 49–50
  with Peppercorns and Brandy, 48–49
  Tournedos of Beef with Béarnaise Sauce, 48

Steak(s) (cont.)
Fillet
    of Beef, Baked, 51
    Strips, Broiled, with Fantastic Sauce,
        46–47
    Tidbits with Yams, 57
Flank, Grilled, 52
Ham, 150
    Baked, with Pineapple, 161
    Baked, with Sherry, 162
    with Orange, 162
    with Peanut Butter and Grapes,
        161
Kebab with Cream Sauce, 56–57
and Kidney
    Pie, 59–60
    Stew, 58–59
Korean Bulgogi (Sirloin Strips), 54–55
Lamb, Broiled, 136
London Broil, Easy, 51–52
Porterhouse, Classic Broiled, 46
Rolls, Stuffed, 54
Sauce, Quick and Easy, 293
Strips with Vegetables, 56
Supreme, 64
Swiss, Braised, 60–61
Steak and Kidney Pie, 59–60
Steak and Kidney Stew, 58–59
Steak in the Blanket, 61
Steak Kebab with Cream Sauce, 56–57
Steak Strips with Vegetables, 56
Steak Supreme, 64
Steiger, Rod, 84
Stew(s)
Beef
    Beaujolais, 83
    with Beer, 82
    Crazy, 84
    Flanken, 83–84
    Old-fashioned, 81
    with Wine, 78–79
Lamb
    and Kidney Bean, 140–141
    Savory, 141
Oxtail, 88–89
Pork and Sauerkraut, 167–168
Steak and Kidney, 58–59
Turkey, with Sherry, 216–217
Veal and Vegetable, 108–109
Stewed Chicken with Dumplings, 197
Stewed Duck with Chestnuts, 226
Storage and refrigeration
canned meats, 7

Storage and refrigeration (cont.)
cured meats, 7
defrosting
    birds, small, 11
    chicken sections, 10
    chops, small, 10
    roasts, 10
    steaks, large, 10
    time, 10–11
    turkeys, large, 10–11
fresh meat, 6–7
frozen meat and fowl, 7
home freezing, 8–9
    "keeping" power, 9
    thawed meat and, 10
leftovers, 8
meat and poultry, 6–11
smoked meats, 7
and tenderizing, 11
thawed meat, 11
Stuffed Crown of Lamb, 133
Stuffed Meatballs—Baked, 66–67
Stuffed Scallopini (Involtini di Vitello),
    102
Stuffed Steak Rolls, 54
Stuffing(s), 271–281
for Beef Scallopini Roulades, 281
Bread, for Poultry, 272
Chestnut, for Poultry, 274–275
    and Brazil Nut, 275
Date and Nut, for Poultry, 276
Fruit
    for Crown Loin of Pork, 279–280
    Nut, and Sausage, for Poultry, 276–
        277
Giblet, for Poultry, 277
Meat
    for Crown of Lamb, 279
    and Sausage, for Crown of Pork, 280
Oyster—Baltimore Style—for Poultry,
    278
Potato, for Poultry, 278
Sage and Onion, for Poultry, 271–272
Savory Kasha, for Poultry, 274
Savory Wild Rice, for Poultry, 273
tips about, 271
Wild Rice, with Grapes for Poultry,
    272–273
Stuffing for Beef Scallopini Roulades,
    281
Succulent Lamb Casserole with Vege-
    tables, 144–145
Suckling Pig, Roasted, 158

Suckling pigs, 148
Swedish Meatballs, 66
Sweet and Sour Barbecued Chicken, 185
Sweetbread and Veal Casserole, 245
Sweetbreads
  and brains, 234–235
    beef, 235
    calf's, 234–235
  Creamed Chicken and, 209
  Maryland Creamed
    No. 1, 244
    No. 2, 244–245
Sweetbreads in Ham Cups, 242–243
Sweetbreads Marinated in Yogurt, 243
Swine, types of, 148
Swiss Steak, Braised, 60–61

Tail of porterhouse, ground, 40–41
T-bone steak, 35
  carving, 25
Tenderizing products, 11
Tenderloin steak, 36
  See also Filet Mignon
Texan Spareribs, 170–171
Tomato Pot Roast, 79
Tongue, 233–234
  Beef, 234
    Basic Fresh, 240
    Fresh, with Vegetable Sauce, 241
    and Ham with Orange, 240–241
  calf's, 234
  lamb, 234
  pork, 234
Tongue and Ham with Orange, 240–241
Tournedos of Beef with Béarnaise
  Sauce, 48
Traxler, Vieri, 102
Tripe, 235
  Fried, 250
Tripe Casserole, 249
Turkey, 181–182, 253
  Breast
    Roast, 216
    Scallopini of, Marsala, 214
  buying tips, 182
  carving, 26–28
  defrosting, 10–11
  facts about, 181–182
  Hash, 217
  Hen, Roast
    No. 1, 212–213
    No. 2, 213

Turkey (*cont.*)
  mature, 179
  Stew with Sherry, 216–217
  Wild, Roast, 256–257
  young tender-meated, 179
Turkey Hash, 217
Turkey Scallopini Rolls (with Prosciutto
  and Swiss Cheese), 215
Turkey Soup, 218
Turkey Stew with Sherry, 216–217
"21 Club," 50

Variety meats, 231–251
  heart, 235
  kidney, 232–233
  liver, 231–232
  recipes, 235–251
  sweetbreads and brains, 234–235
  tongue, 233–234
  tripe, 235
Veal, 90–114
  aging of, 91
  basic cuts of, 92
  and Beef Goulash, 110
  buying, 91–99
    birds, 94
    breast, 99
    center leg roast, 93
    crown roast, 97
    cutlet (or round steak), 94
    cuts (how to identify), 92–99
    foreshank, 98
    Frenched rib chops, 98
    hindsaddle, 93–95
    kidney chops, 96
    kidneys, 233
    loin, 95–96
    loin chops, 96
    loin roast, 95
    neck, 98
    rib, 96–98
    rib chops, 97
    rib roast, 97
    roasts, 93–98
    rolled double sirloin roast, 95
    rolled leg, 93
    rolled loin roast, 96
    scallopini, 94
    shank half of leg, 93–94
    shoulder, 98
    sirloin, 94–95
    standing sirloin roast, 95

Veal
  buying (cont.)
    tips on, 91–92
    what to avoid, 92
  calves, 90
  Casserole, Sweetbread and, 245
  chops
    Frenched rib, 98
    kidney, 96
    loin, 96
    rib, 97
  Dilled, 111–112
  and Green Pepper Casserole, Simple, 112–113
  Loaf, All-American, 113
  quality of, 91
  Quick Creamed, 114
  recipes, 99–114
  Roast, with Ham Stuffing, 108
  Rolls (Saltimbocca), 106
  Sandwiches, 105
  Scallopini
    Breaded, No. 1, 100
    Breaded, No. 2, 101
    Involtini di Vitello, 102
    Marsala, 101–102
    Mozzarella, 103
  and Sour Cream Casserole, 110–111
  types of calves, 90
  and Vegetable Casserole, 109
  and Vegetable Stew, 108–109
Veal and Beef Goulash, 110
Veal and Sour Cream Casserole, 110–111
Veal and Vegetable Casserole, 109
Veal and Vegetable Stew, 108–109
Veal Cutlet in Wine, 100
Veal Hearts and Red Wine Casserole, 248–249
Veal Parmigiana No. 1, 103–104
Veal Parmigiana No. 2, 104

Veal Roast with Ham Stuffing, 108
Veal Sandwiches, 105
Veal Scallopini Mozzarella, 103
Vealers, 90
Venison, 255
  Braised, 266–267
  Chops, Broiled, 268
  Goulash, 267
Venison Goulash, 267
Vinaigrette Sauce, 290

West Indian Chili con Carne, 68–69
Wiener Schnitzel, Old Dutch, 99
Whole ham, 149
Whole leg of lamb, 118
Wild birds and animals, 252–270
  duck, 253–254
  goose, 253
  grouse, 254
  partridge, 254
  quail, 254
  rabbit, 255–256
  recipes, 256–270
  squab, 254
  turkey, 253
  venison, 255
Wild Rice Stuffing, Roasted Guinea Hen with, 229
Wild Rice Stuffing for Poultry, Savory, 273
Wild Rice Stuffing with Grapes for Poultry, 272–273
Wilkens, Emily, 64

Yams, Fillet Tidbits with, 57
Yearling, 116
  mutton, 116–117
Yogurt, Sweetbreads Marinated in, 243